Theory of Laminar Film Condensation

Tetsu Fujii

Theory of Laminar Film Condensation

With 68 Figures

Springer-Verlag

New York Berlin Heidelberg London Paris
Tokyo Hong Kong Barcelona Budapest

Tetsu Fujii
Institute of Advanced Material Study
Kyushu University
Kasuga, Fukuoka 816
Japan

Library of Congress Cataloging-in-Publication Data
Fujii, Tetsu, 1931–
    Theory of laminar film condensation / Tetsu Fujii.
        p.   cm.
    Includes bibliographical references and index.
    ISBN-13:978-1-4612-7816-0   e-ISBN-13:978-1-4612-3152-3
    DOI: 10.1007/978-1-4612-3152-3
    1. Condensation. 2. Vapors. 3. Boundary layer. 4. Laminar flow.
    I. Title.
    QC304.F85   1991
    536'.44 – dc20   91-17241

Printed on acid-free paper.

Camera-ready copy prepared by the author.

9 8 7 6 5 4 3 2 1

ISBN-13:978-1-4612-7816-0

Preface

Since the petroleum crisis in the 1970s, a lot of effort to save energy was made in industry, and remarkable achievements have been made. In the research and development concerning thermal energy, however, it was clarified that one of the most important problems was manufacturing condensing systems with smaller size and higher performance. To solve this problem we need a method which synthesizes selections of the type of condenser, cooling tube and its arrangement, assessment of fouling on the cooling surfaces, consideration of transient characteristics of a condenser, etc.

The majority of effort, however, has been to devise a surface element which enhances the heat transfer coefficient in condensation of a single or multicomponent vapor. Condensation phenomena are complexly affected by a lot of physical property values, and accordingly the results of theoretical research are expressed with several dimensionless parameters. On the other hand, the experimental research is limited to those with some specified cooling surfaces and some specified working fluids. Hence, the basic research of condensation is necessary for criticizing the enhancement effect as well as for an academic interest.

This book deals with steady laminar film condensation on a vertical flat surface as a base for clarifying film condensation phenomena. The numerical analyses are mainly confined to the similarity solution of the vapor-liquid two-phase boundary layer equations, and the results are formulated so as to be useful in understanding the condensation characteristics as well as practical applications. Comparisons between theory and experiment, the method of integral approximation, turbulent film condensation, condensation of a rarefied vapor, fluid flow and heat transfer in a condenser, etc., are not taken up in this book.

This book is composed of the following ten chapters. In Chapter 1 the physical model of condensation is qualitatively explained for the case of a binary vapor mixture. A short history of the development of the water film theory and two-phase boundary layer concept is stated, and some successive research is reviewed. Thereby the significance of the present book will be understood. In Chapter 2 the basic differential equations related to steady laminar film condensation of a binary vapor mixture are listed. In Chapters 3 and 4 the similarity solutions for forced- and free-convection condensation of a binary vapor mixture are presented. A procedure for composing equations of heat and mass transfer characteristics from the boundary values of the solutions and a general way to obtain heat flux and condensation mass flux by combining these equations with phase equilibrium relations are described. In the case of free-convection condensation, it is indicated that the effect of buoyancy force due to temperature distribution in the vapor boundary layer is also considerable as well as due to concentration distribution, especially for a superheated vapor mixture. In Chapter 5 the equations which are obtained in Chapters 3 and 4 are reduced to the case of a pure vapor, and the relevant reported theoretical results are critically summarized. The effects of the superheating of vapor and the convection in the condensate film upon the condensation mass flux are also considered. The latter has never been discussed in previously reported results. In Chapter 6 the effects of various parameters upon heat flux and condensation mass flux in the case of a binary vapor mixture are explained with examples. In Chapters 7 and 8 condensation of a multicomponent vapor mixture is treated by extending the equations obtained in Chapters 3 and 4. In Chapter 9 the representative physical property values in the application of the above-mentioned equations are presented. In Chapter 10 condensation of a pure vapor in the subcritical region is numerically solved. The result gives a limit of applicable range of the above-mentioned equations.

Throughout this book, the forced-convection condensation precedes the free-convection condensation. The reasons are: first, the former mainly takes place in actual condensers, second, the analysis for the former case is simpler than that for the latter case, and third, a relatively smaller number of papers are reported and the problems to be solved have remained for the former case.

Each chapter contains new treatments and new findings. The figures and tables presented will be useful for an intuitive grasp of the characteristics of film condensation and for further improvement in theoretical studies. The formulated results will contribute to the research and development of condenser design. The physical property data used in this book are listed in the Appendix.

Many of the results presented in this book are due to the assistance from the author's academic staffs, particularly Professor H. Uehara, Saga University, Associate Professor Sh. Koyama, Kyushu University, and postgraduate students in the author's laboratory since 1970. In this period the author has benefited greatly from the comments and suggestions offered by Professor J. W. Rose, Queen Mary and Westfield College, U. K. Special acknowledgement is offered to Dr. M. Watabe, Dr. K. Shinzato, Mr. J. B. Lee, Dr. T. Nagata, and Mrs. Yuki Kinoshita for their numerical calculations, preparation of tables and graphs, and manuscript processing.

Nomenclature

$\boldsymbol{A}$	: Eq. (7.3-6), Eq. (7.5-2)
A	: coefficient in Eq. (9.1-2)
a_{kl}	: Eq. (7.2-5)
$\boldsymbol{B}$	: Eq. (7.3-9), Eq. (7.5-3)
B_x	: Eq. (9.1-4)
$\boldsymbol{C}$	: Eq. (7.3-7), Eq. (7.5-9)
$C(H, \mathrm{Pr}_L)$	: Eq. (5.1-11)
$C_F(\mathrm{Pr}_V)$	: function of Pr_V, Eq. (3.4-12)
$C_F(Sc)$	: function of Sc with the same form as Eq. (3.4-12)
C_f	: friction coefficient, Eq. (3.7-4)
C_{f0}	: friction coefficient for single phase convection, Eq. (3.7-5)
$C_G(\mathrm{Pr}_V)$	: function of Pr_V, Eq. (4.4-6)
$C_G(Sc)$	: function of Sc with the same form as Eq. (4.4-6)
c_k	: Eq. (7.2-12)
c_L	: Eq. (4.7-1), Eq. (10.2-4)
c_p	: isobaric specific heat
c_{p12}	: dimensionless isobaric specific heat difference, Eq. (2-8)
c_{pkn}	: dimensionless isobaric specific heat difference, Eq. (7.1-3)
$\boldsymbol{D}$	: Eq. (7.3-14), Eq. (7.5-10)
D	: diffusivity between components 1 and 2 in a binary vapor mixture
D_k^*	: Eq. (8.2-21)
D_{kl}	: coefficient of diffusion, Eq. (7.1-5)
D_{kl}^+	: coefficient of diffusion, Eq. (7.1-4)
D_{kl}^M	: diffusivity of the pair k-l in a multicomponent vapor mixture, Eq. (7.5-1)
D_{kl}^B	: diffusivity between components k and l in a binary vapor mixture, in Chapters 7 and 8
$F_{FL}(\eta_{FL})$	: dimensionless liquid stream function for forced-convection condensation, Eq. (3.1-5)
$F_{FV}(\eta_{FV})$	: dimensionless vapor stream function for forced-convection condensation, Eq. (3.1-6), Eq. (9.2-7)
$F_{GL}(\eta_{GL})$	: dimensionless liquid stream function for free-convection condensation, Eq. (4.1-3)

$F_{GV}(\eta_{GV})$	:	dimensionless vapor stream function for free-convection condensation, Eq. (4.1-4), Eq. (9.2-32)
Fr	:	Froude number $= U_{V\infty}^2/gx$, Eq. (5.5-2)
$\tilde{F}$	:	Eq. (8.2-13)
Ga	:	Galileo number, Eq. (4.1-29)
Gr	:	Grashof number, Eq. (6.2-5), Eq. (8.2-23)
Gr_{Lx}	:	Eq. (10.2-27)
Gr_x	:	Grashof number, Eq. (4.4-7)
g	:	gravitational acceleration
H	:	phase change number $= c_{p_L}(T_s - T_w)/\Delta h_V$, Eq. (5.1-2)
H_i	:	phase change number $= c_{p_L}(T_i - T_w)/\Delta h_V$, Eq. (3.1-33)
h	:	enthalpy, or height of control surface in Fig. 2-2
j_k	:	diffusion mass flux of component k (=1, 2), Eq. (8.2-27)
j_k^*	:	Eq. (8.2-19)
K	:	Eq. (5.2-9), Eq. (9.1-8)
k	:	coefficient in Eq. (4.4-12), Eq. (6.1-8)
ℓ	:	length of cooling surface, index in Eq. (4.4-12)
M	:	apparent molecular weight of mixture, Eq. (4-2)
M_k	:	molecular weight of component k (=1, 2, ..., n)
$\dot{M}_{FL}$	:	dimensionless condensate mass flux defined by physical properties of condensate for forced-convection condensation

$$= \sum_{k=1}^{n} \dot{M}_{kFL} = \frac{\dot{m}_x x}{\mu_L \mathrm{Re}_{Lx}^{1/2}} = \frac{F_{FLi}}{2},$$

Eq. (3.1-32), Eq. (7.2-11)

$\dot{M}_{FV}$	:	dimensionless condensate mass flux defined by physical properties of vapor mixture for forced-convection condensation $= R\dot{M}_{FL} = F_{FVi}/2$, Eq. (3.4-9)
$\dot{M}_{GL}$	:	dimensionless condensate mass flux defined by physical properties of condensate for free-convection condensation

$$= \sum_{k=1}^{n} \dot{M}_{kGL} = \left(\frac{\dot{m}_x x}{\mu_L}\right)\left(\frac{Ga_x}{4}\right)^{-\frac{1}{4}} = 3R^{-1}F_{GVi}, \quad \text{Eq. (4.1-28)}$$

$\dot{M}_{GV}$	:	dimensionless condensate mass flux defined by physical properties of vapor mixture for free-convection condensation $= \omega_i^{-\frac{1}{4}} R\dot{M}_{GL}$, Eq. (4.4-10a), Eq. (4.4-10b)
$\dot{M}_{kFL}$	:	component k of $\dot{M}_{FL}$, Eq. (3.1-31), Eq. (7.2-10)
$\dot{M}_{kGL}$	:	component k of $\dot{M}_{GL}$, Eq. (4.1-27), Eq. (8.1-7)
$(\dot{M}_{FV})_{sat/sup}$	:	Eq. (5.2-7)
m	:	Eq. (3.5-7)
$\dot{m}$	:	condensation mass flux (mass condensed per unit time per unit area)

$\dot{m}_k$: condensation mass flux of component k

N_k : Eq. (8.2-18)

n : Eq. (4.5-6), exponent in Eq. (4.5-5)

Nu_c : Nusselt number corresponding to α_c, Eq. (3.4-6) and Eq. (4.4-3)

Nu_{Li} : Nusselt number corresponding to α_i, Eq. (3.4-5) and Eq. (4.4-2)

Nu_{Lw} : Nusselt number corresponding to α_w, Eq. (3.4-4) and Eq. (4.4-1)

Nu^* : modified Nusselt number, Eq. (5.7-18)

$\overline{Nu^*}$: modified average Nusselt number, Eq. (5.7-20)

$\mathbf{P}$: Eq. (7.3-10), Eq. (7.5-5)

$\mathbf{P}^{-1}$: Eq. (7.3-11), Eq. (7.5-6)

Pr : Prandtl number $= \mu c_p / \lambda$

p : static pressure

p_c : critical pressure

p_k : partial pressure of component k ($=1, 2, \ldots, n$)

p_{kl} : Eq. (7.3-10)

q_c : heat flux in the vapor side at the vapor-liquid interface

q_i : heat flux in the condensate side at the vapor-liquid interface

q_{kl} : Eq. (7.3-11)

q_w : local heat flux at cooling surface

q_0 : uniform heat flux at cooling surface, Section 5.7

$(q_{wx})_{sat/sup}$: Eq. (5.2-12)

R : $\rho\mu$ ratio $=(\rho_L \mu_L / \rho_V \mu_V)^{1/2}$, Eq. (3.1-30)

$\Re$: universal gas constant

r_T : dimensionless temperature corresponding to representative physical properties, Eq. (9.1-16), Eq. (9.1-20)

Re_V : vapor Reynolds number, Eq. (3.4-7)

Re_L : two-phase Reynolds number $= U_{V\infty} x / \nu_L$, Eq. (3.1-34)

Re^* : film Reynolds number, Eq. (5.7-19) and Eq. (5.7-21)

Sc : vapor Schmidt number $= \nu_V / D$

Sc_k : inverse of eigenvalue of matrix $\mathbf{A}$

Sc_{kl} : Eq. (7.2-4)

Sh : Sherwood number, Eq. (3.5-2) and Eq. (4.5-1)

Sh^*_{kx} : Eq. (8.2-20)

T : temperature

T_c : critical temperature

T_{Lr}, T_{Vr} : temperature for representative physical properties

T_s : saturation temperature or dew point

T^* : temperature relative to the cooling surface temperature T_w

T_i^0, T_w^0 : guessed initial temperature in numerical calculation

U : velocity component in the x direction

$U_{V\infty}$: U in the bulk

V : velocity component in the y direction

W_{kV} : mass concentration (mass fraction) of component k in the vapor phase, Eq. (2-9), Eq. (7.1-2)

W_{kL} : mass concentration of component k in liquid phase, Eqs. (2-28), (3.1-35), (4.1-30), (7.1-10), and (7.2-13)

W_{kR} : Eq. (7.3-26), Eq. (7.5-13), Eq. (8.1-10)

W_R : $(W_{1Vi} - W_{1L})/(W_{1V\infty} - W_{1L})$, Eq. (3.1-37)

W_k^* : Eq. (8.2-11)

w_L : liquid phase line or boiling point line in the phase equilibrium diagram , Eq. (2-29)

w_V : vapor phase line or dew point line in the phase equilibrium diagram, Eq. (2-30)

X_{kL}, X_{kV} : molar concentration of component k, Eq. (7.1-11)

x : coordinate measuring distance along the plate from the leading edge, Fig. 2-1

x_c : boundary between forced-convection and free-convection condensation regions, Eq. (5.5-3)

x_{cm} : apparent boundary between forced-convection and free-convection condensations in an average heat transfer coefficient, Eq. (5.5-5)

x_r : x coordinate where the local Nusselt number is equal to the average Nusselt number

Y_{F1}, Y_{G1} : left-hand side term of Eq. (5.2-1) or Eq. (5.4-1)

Y_{F2}, Y_{G2} : the first term on the right-hand side of Eq. (5.2-1) or Eq. (5.4-1)

Y_{F3}, Y_{G3} : the second term on the right-hand side of Eq. (5.2-1) or Eq. (5.4-1)

y : coordinate measuring distance normal to the cooling surface, Fig. 2-1

α_c : convective heat transfer coefficient of vapor mixture, Eq. (3.4-3)

α_i : convective heat transfer coefficient of condensate at the vapor-liquid interface, Eq. (3.4-2)

α_w : convective heat transfer coefficient of condensate at the cooling surface, Eq. (3.4-1)

β : mass transfer coefficient, Eq. (3.5-1)

β_{kx}, β_{kx}^* : mass transfer coefficients corresponding to j_k and j_k^*, Eqs. (8.2-27) and (8.2-19), respectively

γ_k : activity constant of component k

Δ	:	vapor boundary layer thickness, Fig. 2-1
Δh_V	:	latent heat of condensation
$\Delta T_{V\infty}$	:	degrees of superheat of bulk vapor $= T_{V\infty} - T_{s\infty}$
δ	:	condensate film thickness, Fig. 2-1
ε	:	Eq. (7.5-7), Eq. (10.1-24)
ζ	:	Eq. (7.5-8)
η_{FL}	:	similarity variable for condensate film in forced-convection condensation, Eq. (3.1-3), Eq. (10.1-7)
η_{FV}	:	similarity variable for the vapor boundary layer in forced-convection condensation, Eq. (3.1-4), Eq. (9.2-6)
η_{GL}	:	similarity variable for condensate film in free-convection condensation, Eq. (4.1-1), Eq. (10.2-2)
η_{GV}	:	similarity variable for the vapor boundary layer in free-convection condensation, Eq. (4.1-2), Eq. (10.2-31)
$\tilde{\eta}$	:	Eq. (8.2-12)
Θ_F, Θ_G	:	$(T - T_w)/(T_{V\infty} - T_w)$
$\Theta_{FL}(\eta_{FL})$	:	dimensionless condensate temperature in forced-convection condensation, Eq. (3.1-7)
$\Theta_{FV}(\eta_{FV})$	:	dimensionless vapor temperature in forced-convection condensation, Eq. (3.1-8)
$\Theta_{GL}(\eta_{GL})$	:	dimensionless condensate temperature in free-convection condensation, same form as Eq. (3.1-7)
$\Theta_{GV}(\eta_{GV})$	:	dimensionless vapor temperature in free-convection condensation, same form as Eq. (3.1-8)
$\Theta_{sFV}, \Theta_{sGV}$	:	dimensionless saturation temperature, $= (T_{V\infty} - T_s)/(T_{V\infty} - T_i)$
κ	:	thermal diffusivity
λ	:	thermal conductivity
μ	:	dynamic viscosity
ν	:	kinematic viscosity
ξ	:	Eq. (5.4-6)
ρ	:	density
τ	:	skin friction (shear stress), Eqs. (3.7-1) and (3.7-2)
Φ_F	:	Eq. (7.3-5), Eq. (7.3-13)
Φ_F^*	:	Eq. (7.3-12)
$\Phi_F(\eta_{FV})$	:	normalized concentration of vapor in forced-convection condensation, Eq. (3.1-9)
$\Phi_G(\eta_{GV})$	:	normalized concentration of vapor in free-convection condensation, same form as Eq. (3.1-9)
Φ_{kF}	:	Eq. (7.2-1)
Φ_{kF}^*, Φ_{kG}^*	:	Eq. (7.3-23), Eq. (8.1-3)
ϕ	:	relative humidity
ϕ_i	:	Eq. (7.4-3)

χ	:	function of Pr_L/RH, Eq. (5.1-9)
χ_i, χ_{ki}	:	Eq. (4.5-3), Eq. (8.1-13), Eq. (8.2-25)
χ_i^+	:	Eq. (4.4-9), Eq. (8.1-14), Eq. (8.2-34)
$\Psi_L(x,y)$	:	liquid stream function, Eq. (3.1-1), Eq. (10.1-6)
$\Psi_V(x,y)$	:	vapor stream function, Eq. (3.1-2), Eq. (9.2-5)
Ω_T	:	Eq. (4-5), Eq. (8-2)
Ω_W, Ω_{Wk}	:	Eq. (4-4) and Eq. (8-3), respectively
ω	:	Eq. (6.2-6)
ω_i	:	Eq. (4.4-8), Eq. (8.1-15)
ω_s	:	Eq. (8.2-14)
ω_T	:	Eq. (4.1-11), Eq. (8.1-4)
ω_W	:	Eq. (4.1-10)
ω_k^*	:	Eq. (8.1-5), Eq. (8.2-8), Eq. (8.2-9)

Superscripts

prime	:	differential derivative with respect to η_L or η_V
bar	:	average over $x=0$ to ℓ
o	:	reference condition
*	:	values after orthogonal transformation

Subscripts

o	:	reference condition
1, 2	:	volatile and less volatile components of binary vapor mixture, respectively
cal	:	calculated by the algebraic equations
e	:	values at T_e or W_{1Ve}
exp	:	experimental values
F	:	forced-convection condensation
film	:	based on stagnant film theory
FU	:	corresponding to Eq. (5.1-8)
G	:	free-convection condensation
i	:	vapor-liquid interface
k	:	component k (numbered in order of low boiling point)
k, l, m, n	:	component of vapor mixture
L	:	condensate
ℓ	:	at $x = \ell$
Nu	:	corresponding to Eq. (5.3-2)
r	:	representative value for physical properties
s, sat	:	saturated

sup	:	superheated
sim	:	similarity solution
V	:	vapor
w	:	wall surface
x	:	local values at x
∞	:	bulk, at $y = \infty$

Contents

1

Introduction

A vapor condenses when it contacts a solid wall (cooling surface) with a temperature lower than the saturation temperature of the vapor or the dew point in the case of a multicomponent vapor mixture. The main objective of condensation heat transfer research in this situation is to quantitatively clarify various effects upon the condensation mass flux and the heat flux at the cooling surface, to which the heat is transferred through condensate. The condensate becomes film or dropwise depending on whether the cooling surface is wettable or not. Steady laminar film condensation is treated in this book.

The cooling surface is covered by the condensate film and outside the film a boundary layer of vapor or vapor mixture is formed. The film thickness, the boundary layer thickness, the thermodynamic state at the vapor-liquid interface, the distributions of temperature and mass concentration in the direction normal to the surface, etc., are strongly affected by the velocity of main vapor flow. When the vapor velocity is very low, these are instead determined by a buoyancy force due to the density difference between the condensate and vapor. The former is named forced-convection condensation and the latter is named free-convection condensation. Condensation between these two condensation regions is called combined forced- and free-convection condensation.

In this book the basic characteristics of laminar film condensation based on the results of theoretical analysis of condensation of a miscible binary vapor mixture on a flat surface is systematically described. First, the mechanism of condensation is qualitatively explained. Figure 1-1(a) schematically shows the transverse (y direction) distributions of temperature T and mass concentration $W_{kV}(k = 1, 2)$ in the condensate film and the vapor boundary layer in the case of steady condensation of a binary vapor mixture. These variations of T and W_{kV} on a diagram of phase equilibrium are shown in Fig. 1-1(b). When a binary vapor mixture is cooled in contact with a solid wall, the less volatile component ($k = 2$) condenses more than the volatile component ($k = 1$) according to the phase equilibrium rule. Consequently the volatile component becomes dense at the vapor-liquid interface in the vapor phase because the sum of mass concentration is kept constant in the vapor boundary layer. Since a binary vapor mixture with a constant mass concentration in the bulk is steadily supplied to the vapor-liquid interface, the above trend is accelerated. On the other

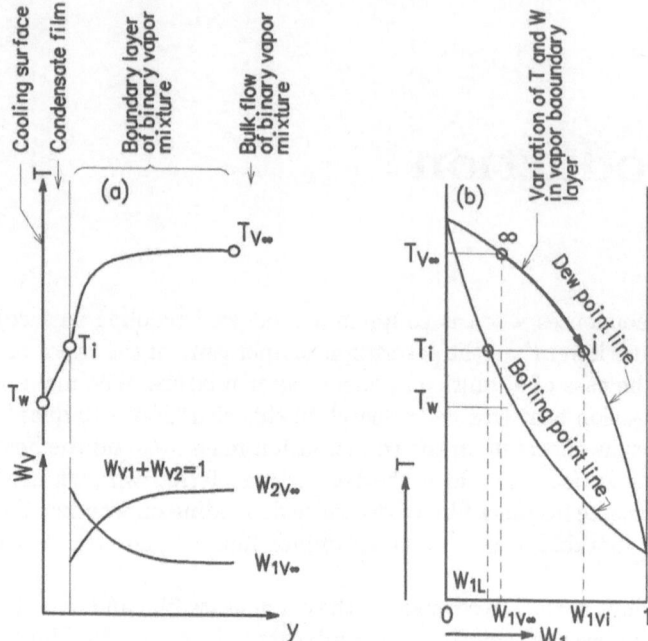

FIGURE 1-1.   Temperature T and mass concentration W in condensation. Subscripts, ∞: bulk; i: vapor-liquid interface; w: cooling wall surface; 1: volatile component; 2: less volatile component; L: condensate. (a) Distributions of T and W in the condensate film and the vapor boundary layer. The y denotes normal distance from the surface. (b) Variation of T and W on a diagram of phase equilibrium.

hand, convective mass transfer acts so that the mass concentration at the vapor-liquid interface approaches that in the bulk. A balance of these two actions makes a steady distribution of mass concentration, as already shown in Fig. 1-1(a). At the same time, the temperature of the vapor mixture, which satisfies the phase equilibrium rule, decreases from the bulk to the vapor-liquid interface. The temperature in the vapor and liquid phases can be considered to be continuous at the vapor-liquid interface, except in the case of a rarefied vapor. Since the mechanism of temperature drop in the condensate film (nearly conductive heat transfer) is the same as in the case of condensation of a pure vapor, the heat transfer rate and condensation rate for a vapor mixture become smaller than those for a pure vapor in the case of same temperature difference between the cooling surface and the bulk vapor. A temperature drop in the vapor boundary layer of a superheated pure vapor takes place by the same mechanism as the single phase convective heat transfer with suction, because the temperature of the vapor-liquid interface is considered to be the saturation temperature.

The most effective means at present to quantitatively analyze the relevant condensation problem is the two-phase boundary layer theory. Its establishment and the successive research are briefly described below.

In 1861 Joule[1] performed an experiment of condensation of steam inside a vertical tube and substantiated the existence of thermal resistance in the condensate film, which had been under debate, and obtained the data for the design of a surface condenser. Incidentally, the first steam surface condenser began its operation in 1860 for a marine engine. In 1873 Reynolds[2] performed an experiment of condensation of steam with air to test the possibility of preventing water hammer due to condensation in the cylinder of a steam engine. He obtained the result that the condensation rate is drastically reduced by mixing a small amount of air although the condensation rate of a pure vapor is infinitely large. In modern steam surface condensers the thermal resistance of condensate film is predominant for most cooling tubes, and that of vapor layer is predominant for cooling tubes in an air cooling zone.

In 1916 Nusselt[3] proposed the water film theory for condensation of otherwise quiesent pure steam on a vertical flat surface and on a horizontal cylindrical surface, and proved his theory by comparing the condensate film thickness predicted with that measured by English and Donkin.[4] In the same paper he also discussed the effects of vapor superheating, presence of air, vapor velocity, and condensate inundation upon condensation heat transfer. However, these results are scarcely referred at present.

After the Nusselt theory was introduced to America by Monrad and Badger[5] in 1930, it was applied to data processing in all kinds of experiments of condensation heat transfer, and proved to be somewhat useful even in the case of the condensation of an organic substance inside a horizontal tube. This success came from the fact that Nusselt's equation contains the phase change number, which is one of the dimensionless numbers dominating condensation phenomenon.

Though a lot of effort was spent in the design of a steam condenser with increase in size or capacity, it was revealed that the condensation heat transfer coefficient measured in actual condensers never coincides with that predicted by the Nusselt theory. Consequently, there arose the question: is the Nusselt theory itself is exact or not? In this context Sparrow and Gregg,[6] in 1959, solved exactly the conservation equations of momentum and energy for the condensate film in free-convection condensation of a pure vapor by introducing the similarity transformation which had been used in the boundary layer treatment of single phase free convection.

In 1960 Cess[7] did an analysis of forced-convection condensation on a horizontal flat surface. In this analysis a numerical solution for the wall shear stress in the case of the boundary layer with suction was ingeniously applied. In 1961 Koh et al.[8] numerically solved a simultaneous conservation equation of momentum and energy of liquid and momentum of vapor in the free-convection condensation of a pure vapor. In this paper the concept of a two-phase boundary layer is clearly described. The concept provided its effectiveness in successive research, especially in the analysis of condensation of a multicomponent vapor.

Table 1-1 shows literature since 1959 concerning theoretical studies of condensation on a flat surface. Most of these results were achieved with the aid of developments in electronic computers and numerical calculation techniques. Though a great many numerical results have been accumulated in the form of tables and graphs, most of the conclusions in the literature are qualitative and lack generality. For example, the heat transfer rate in condensation of a specified binary or ternary vapor mixture is expressed as a ratio to that of a corresponding imaginary pure vapor (with the same saturation temperature as the bulk temperature and with the physical properties of the bulk vapor mixture) for a specified condensation condition, or is expressed in comparison with the Nusselt theory even in the case of forced-convection condensation. Fujii and Uehara's[9] equation in no. 20 of Table 1-1, which was obtained by formulating numerical results for forced-convection condensation, is comparable to the Nusselt equation for free-convection condensation. In the papers of nos. 30, 31, 33, and 37–42, the idea of formulation of numerical results are applied to the cases of binary and multicomponent vapors.

In this book it is explained that two couples of simultaneous algebraic equations of heat transfer and mass transfer for both forced- and free-convection condensation of a binary vapor mixture are most fundamental in the prediction of condensation characteristics, i.e., the equations involve not only the characteristics of condensation of a pure vapor but also enable us to treat the case of a multicomponent vapor mixture. By using these equations, therefore, we can accurately derive all the results of the literature shown in Table 1-1 .

TABLE 1-1. Theoretical studies of steady laminar film condensation on a flat surface.

No.	Year	Authors	Mode & Method	Substances	Ref. No. cited
1	1959	Sparrow & Gregg	G, Similarity (Condensate film)	pure vapor	106, 511
2	1960	Cess	F, Analogy	pure vapor	107, 503
3	1961	Sparrow & Eckert	G, Similarity	air + water, superheated	515, 609
4	1961	Koh et al.	G, Similarity	pure vapor	108, 514
5	1961	Koh	G, Integral	pure vapor	—
6	1961	Chen	G, Integral & Perturbation	pure vapor	512
7	1962	Koh	F, Similarity	binary, superheated	202, 507, 614
8	1962	Koh	F, Similarity	pure vapor	505
9	1964	Stewart & Prober	F, Film theory & Matrix transformation	multicomponent	702, 807
10	1964	Sparrow & Lin	G, Similarity	air + water	610
11	1966	Shekriladze & Gomerauri	F, G, Approximate	pure vapor	518, 908
12	1966	Jacobs	F + G, Integral	pure vapor	517
13	1966	Minkowycz & Sparrow	G, Local similarity	air + water	203, 516, 611, 910
14	1967	Sparrow et al.	F, Similarity & Integral	air + water	504, 601, 906
15	1969	Minkowycz & Sparrow	F, Similarity	air + water superheated	509, 602
16	1969	Sparrow & Marshall	G, Similarity	$CH_3OH + H_2O$	617
17	1969	Rose	G, Integral	air + water	—
18	1969	Denny & Mills	F + G, Finite difference	pure vapor	907
19	1971	Denny et al.	F + G, Finite difference	air + water	621
20	1972	Fujii & Uehara	F + G, Approximate & Correlation	pure vapor	109, 502
21	1972	Denny & Jusionis	F + G, Finite difference	air + (H_2O,NH_3, C_2H_5OH, C_4H_9OH, CCl_4, CCl_2F_2)	622

F:Forced-convection condensation, G: Free-convection condensation.
The first figure of Ref. No. cited denotes chapter.

TABLE 1-1.   (continued)

No.	Year	Authors	Mode & Method	Substances	Ref. No. cited
22	1972	Denny & Jusionis	F + G, Finite difference	$CH_3OH + H_2O$, $C_3H_7OH + H_2O$, $(CH_3)CO + H_2O$, $(CH_3)CO + CCl_4$	623
23	1972	Mori & Hijikata	G, Integral	air + water	—
24	1972	Hijikata & Mori	F, Integral	air+water,air+KCl	608
25	1972	Fujii et al.	G, Approximate	pure vapor	519
26	1973	Tamir	G, Integral	$CH_3OH + H_2O$	
27	1974	Taitel & Tamir	G, Similarity & Integral	air+CH_3OH+H_2O, $(CH_3)CO$+CH_3OH+H_2O	804
28	1976	Lucas	F + G, Finite difference & Integral	$CH_3OH + H_2O$	624
29	1976	Sage & Estrin	G, Similarity	$Ne + N_2 + H_2O$, N_2+CCl_2F_2+H_2O, N_2+CH_4+H_2O	805
30	1977	Fujii et al.	F, Similarity & Correlation	gas + vapor	301, 603
31	1978	Fujii et al.	G, Similarity & Correlation	gas + vapor	401, 612
32	1980	Stephan & Laesecke	F, G, F + G, Modified film theory	$CH_3OH + H_2O$, air + H_2O	625
33	1980	Rose	F, Correlation	gas + vapor	604
34	1980	Hijikata et al.	G, Integral	air+H_2O,air+$C_2Cl_3F_3$	619
35	1980	Fujii & Kato	F + G, Correlation	binary vapor	625
36	1985	Kotake	G, Perturbation	multicomponent	—
37	1984	Fujii & Koyama	F, Similarity & Correlation	air+CH_3OH+H_2O	703
38	1986	Koyama et al.	G, Similarity & Correlation	$C_2H_5OH + H_2O$, air + H_2O	—
39	1987	Fujii et al.	F, Similarity & Correlation	C_2H_5OH+H_2O, $C2H_2F4$+CCl_3F, air + H_2O, air + C_2H_5OH, $CHClF_2$+$C_2Cl_2F_4$, air+$C_2Cl_2F_4$,air+$CHClF_2$	303, 615
40	1987	Koyama et al.	F, Matrix transformation	multicomponent	704
41	1987	Koyama et al.	G, Matrix transformation	multicomponent	801
42	1989	Fujii et al.	G, Similarity	CH_3OH+C_2H_5OH+H_2O	803

REFERENCES

1. Joule, J. P., On the Surface Condensation of Steam, *Proc. Roy. Soc. London*, 151, 133–160 (1861).

2. Reynolds, O., On the Condensation of a Mixture of Air and Steam upon Cold Surface, *Proc. Roy. Soc. London*, 21, 14, 275–281 (1873).

3. Nusselt, W., Die Oberflächenkondensation des Wasserdampfes, *Zeit. VDI*, 60, 27, 541–546, ibid., 28, 569–575 (1916).

4. English, T. and B. Donkin, Transmission of Heat from Surface Condensation through Metal Cylinders, *Proc. Inst. Mech. Engr.*, 501–533 (1896).

5. Monrad, C. C. and W. L. Badger, The Condensation of Vapors, *Ind. Eng. Chem.*, 22, 1, 1103–1112 (1930).

6. Sparrow, E. M. and J. L. Gregg, A Boundary-Layer Treatment of Laminar-Film Condensation, *Trans. ASME, J. Heat Transfer*, 81, 13–18 (1959).

7. Cess, R. D., Laminar-Film Condensation on a Flat Plate in the Absence of a Body Force, *Z. Angew. Math. Phys.*, XI, 426–433 (1960).

8. Koh, J. C. Y., E. M. Sparrow, and J. P. Hartnett, The Two Phase Boundary Layer in Laminar Film Condensation, *Int. J. Heat Mass Transfer*, 2, 69–82 (1961).

9. Fujii, T. and H. Uehara, Laminar Filmwise Condensation on a Vertical Surface, *Int. J. Heat Mass Transfer*, 15, 217–233 (1972).

2

Basic Equations for Laminar Film Condensation of a Binary Vapor

The basic differential equations for steady laminar film condensation of a binary vapor mixture on a vertical flat cooling surface, for which the vapor flows parallel and downwards, are concisely explained in this chapter. The details of the derivation of the equations are described in Refs. 1, 2, and 3.

Figure 2-1 shows the physical model of the vapor-liquid two-phase boundary layer and the coordinate system, where x is the distance measured along the surface from the leading edge to the direction of liquid and vapor flow, y is the normal distance from the surface, δ is the condensate film thickness, Δ is the vapor boundary layer thickness, U and V are the velocity components of the x and y direction, respectively, T is the temperature, W is the mass concentration (or mass fraction) of a component of the vapor mixture, g is the gravitational acceleration, and p is the static pressure. The subscripts L and V denote the values in the condensate film and vapor phase, respectively, the subscripts w, i, and ∞ at the cooling surface, vapor-liquid interface, and bulk vapor, respectively, and the subscript 1 the component with lower boiling points.

For simplicity, we consider the case where the following assumptions necessary to derive the similarity solution (see Sections 3.1 and 4.1) are valid, and thereby clarify principal features of laminar film condensation:

1. T_w, T_i, $T_{V\infty}$, W_{1Vi}, $W_{1V\infty}$, $U_{Vi}(=U_{Li})$, and $U_{V\infty}$ are independent of x.

2. Both the condensate film and the vapor boundary layer develop from the leading edge of the cooling surface. This assumption is not valid in the case of free-convection condensation where the molecular weight of the more volatile component is smaller than that of the less volatile component (see Chapter 4 and Section 6.4).

3. The phase equilibrium condition is satisfied at the vapor-liquid interface and condensation takes place only there. See Sections 3.3.3 and 4.3.3 for the possibility of mist or fog formation in the vapor boundary layer.

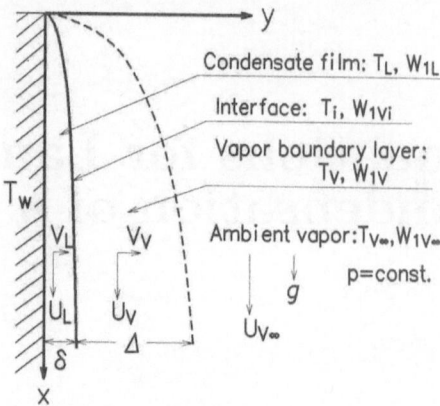

FIGURE 2-1.   Physical model and coordinate system for condensation of a binary vapor mixture. (This figure is extremely enlarged in the y direction. Actual dimensions of δ and Δ are of the orders of 0.1 and 1–100 mm, respectively.)

4. Temperature and velocity are continuous at the vapor-liquid interface. This assumption is not valid for a rarefied vapor.

5. The condensate is miscible.

6. The variation of physical properties with concentration and temperature is considered only for buoyancy force terms, and proper representative values can be used for the other physical quantities (see Chapter 9).

7. Density of the condensate is much larger than that of the vapor mixture. This assumption is not valid at the state near the critical point (see Chapter 10).

8. The effects of thermal diffusion and diffusion-thermo can be ignored (see Ref. 3).

9. The vapor mixture can be treated as an ideal gas.

Using the above assumptions and the boundary layer concept, we can derive the following basic equations of conservation of total mass, momentum, energy, and mass of component 1 for a binary vapor mixture.

For the condensate film:

$$\frac{\partial U_L}{\partial x}+\frac{\partial V_L}{\partial y} = 0 \qquad\qquad \text{(continuity)} , \quad (2\text{-}1)$$

$$U_L\frac{\partial U_L}{\partial x}+V_L\frac{\partial U_L}{\partial y} = \nu_L\frac{\partial^2 U_L}{\partial y^2}+g \qquad\qquad \text{(momentum)} , \quad (2\text{-}2)$$

$$U_L\frac{\partial T_L}{\partial x}+V_L\frac{\partial T_L}{\partial y} = \kappa_L\frac{\partial^2 T_L}{\partial y^2} \qquad\qquad \text{(energy)} . \quad (2\text{-}3)$$

For the vapor boundary layer:

$$\frac{\partial U_V}{\partial x} + \frac{\partial V_V}{\partial y} = 0 \qquad \text{(continuity)}, \quad (2\text{-}4)$$

$$U_V \frac{\partial U_V}{\partial x} + V_V \frac{\partial U_V}{\partial y} = \nu_V \frac{\partial^2 U_V}{\partial y^2} + g\left(1 - \frac{\rho_{V\infty}}{\rho_V}\right) \qquad \text{(momentum)}, \quad (2\text{-}5)$$

$$U_V \frac{\partial T_V}{\partial x} + V_V \frac{\partial T_V}{\partial y} = \kappa_V \frac{\partial^2 T_V}{\partial y^2} + D\, c_{p12} \frac{\partial W_{1V}}{\partial y} \frac{\partial T_V}{\partial y} \qquad \text{(energy)}, \quad (2\text{-}6)$$

$$U_V \frac{\partial W_{1V}}{\partial x} + V_V \frac{\partial W_{1V}}{\partial y} = D \frac{\partial^2 W_{1V}}{\partial y^2} \qquad \text{(continuity of component 1)}, \quad (2\text{-}7)$$

where ν is the kinematic viscosity, κ is the thermal diffusivity, ρ is the density, and D is the diffusivity between components 1 and 2.

The dimensionless isobaric specific heat difference c_{p12} in the second term (diffusion term) on the right-hand side of Eq. (2-6) is given as

$$c_{p12} = \frac{c_{p1V} - c_{p2V}}{c_{p1V} W_{1V} + c_{p2V} W_{2V}} = \frac{c_{p1V} - c_{p2V}}{c_{pV}}, \qquad (2\text{-}8)$$

where c_p is the isobaric specific heat. Though the variation of c_{p12} in the vapor boundary layer is larger than that of the other physical properties, c_{p12} is taken to be uniform because the second term on the right-hand side of Eq. (2-6) is much smaller than the first term.

The mass concentrations W_{1V} and W_{2V} in the vapor phase are defined by

$$W_{1V} = \frac{\rho_{1V}}{\rho_V}, \qquad (2\text{-}9a)$$

$$W_{2V} = \frac{\rho_{2V}}{\rho_V}, \qquad (2\text{-}9b)$$

where

$$\rho_V = \rho_{1V} + \rho_{2V}. \qquad (2\text{-}10)$$

Hence, the relation between W_{1V} and W_{2V} is expressed as

$$W_{1V} + W_{2V} = 1. \qquad (2\text{-}11)$$

From this relation it is clear that the conservation equation of either W_{1V} or W_{2V} is sufficient in the basic equation system. In other words, Eq. (2-4) already contains the conservation of W_{1V} and W_{2V} in physical meaning (see Ref. 1). The partial pressures p_1 and p_2, and W_{1V} and W_{2V} can be mutually changed by the equations

$$\frac{p_1}{p} = \left(1 + \frac{M_1 W_{2V}}{M_2 W_{1V}}\right)^{-1}, \qquad \frac{p_2}{p} = \left(1 + \frac{M_2 W_{1V}}{M_1 W_{2V}}\right)^{-1}, \qquad (2\text{-}12a,b)$$

$$W_{1V} = \left(1 + \frac{M_2 p_2}{M_1 p_1}\right)^{-1}, \qquad W_{2V} = \left(1 + \frac{M_1 p_1}{M_2 p_2}\right)^{-1}, \qquad (2\text{-}13a,b)$$

where M_1 and M_2 are the molecular weights of components 1 and 2, respectively.

The boundary conditions for the basic equations are written as

at $y = 0$:

$$U_L = 0 , \tag{2-14}$$
$$V_L = 0 , \tag{2-15}$$
$$T_L = T_w ; \tag{2-16}$$

as $y \to \infty$:

$$U_V = U_{V\infty} , \tag{2-17}$$
$$T_V = T_{V\infty} \quad (= T_{s\infty} + \Delta T_{V\infty}) , \tag{2-18}$$
$$W_{1V} = W_{1V\infty} ; \tag{2-19}$$

and the compatibility conditions at the vapor-liquid interface are written as

at $y = \delta$:

$$U_{Li} = U_{Vi} = U_i , \tag{2-20}$$

$$\left(\mu_L \frac{\partial U_L}{\partial y} \right)_i = \left(\mu_V \frac{\partial U_V}{\partial y} \right)_i , \tag{2-21}$$

$$\left\{ \rho_L \left(U_L \frac{d\delta}{dx} - V_L \right) \right\}_i = \left\{ \rho_V \left(U_V \frac{d\delta}{dx} - V_V \right) \right\}_i \tag{2-22}$$

$$= \dot{m}_x = \dot{m}_{1x} + \dot{m}_{2x} , \tag{2-23}$$

$$T_{Li} = T_{Vi} = T_i , \tag{2-24}$$

$$\left(\lambda_L \frac{\partial T_L}{\partial y} \right)_i = \Delta h_V \dot{m}_x + \left(\lambda_V \frac{\partial T_V}{\partial y} \right)_i , \tag{2-25}$$

$$W_{1V} = W_{1Vi} , \tag{2-26}$$

$$\left(\rho_V D \frac{\partial W_{1V}}{\partial y} \right)_i = \dot{m}_{1x} - W_{1Vi} \dot{m}_x , \tag{2-27}$$

where μ is the dynamic viscosity, λ is the thermal conductivity, Δh_V is the latent heat of condensation, $\dot{m}$ is the condensation mass flux, and the subscript x denotes the local value at x.

The mass concentration of component 1 in the condensate film W_{1L} is expressed as

$$W_{1L} = \frac{\dot{m}_{1x}}{\dot{m}_{1x} + \dot{m}_{2x}} = \frac{\dot{m}_{1x}}{\dot{m}_x} \tag{2-28}$$

and the relations of phase equilibrium at the vapor-liquid interface are expressed as

$$W_{1Li} = w_L(T_i, p) \quad \text{(saturated liquid or boiling point)} , \tag{2-29}$$
$$W_{1Vi} = w_V(T_i, p) \quad \text{(saturated vapor or dew point)} . \tag{2-30}$$

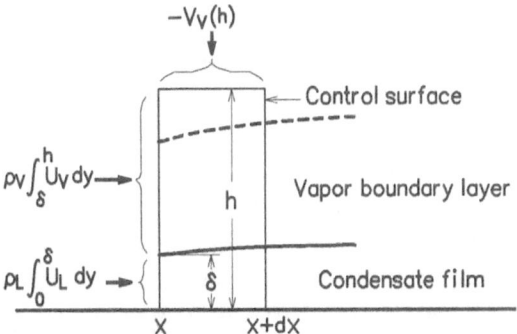

FIGURE 2-2. Control surface in the section $[x, x + dx]$.

Some examples for the functional forms of w_L and w_V are shown in the Appendix, Fig. 6.3-1(a), and Figs. 6.3-2(a) and (b).

We now explain the derivation of the compatibility conditions mentioned above.

The continuity conditions of the velocity component U and the shearing stress at the vapor-liquid interface, Eqs. (2-20) and (2-21), respectively, can be derived based on the fact that the value of $d\delta/dx$ is much smaller than unity.

The continuity conditions of condensation mass flux [Eq. (2-22)] can be derived in the following manner. In the section $[x, x+dx]$ shown in Fig. 2-2, we consider a control surface with height h, which is sufficiently larger than the vapor boundary layer thickness. The mass flow rates of condensate at x and $x + dx$ are given, respectively by

$$\rho_L \int_0^\delta U_L dy \,, \quad \rho_L \left\{ \int_0^\delta U_L dy + \frac{d}{dx}\left(\int_0^\delta U_L dy \right) dx \right\} . \qquad \text{(2-31a,b)}$$

Therefore the increase in the mass flow rate of condensate towards the x direction within the section dx is

$$\rho_L \frac{d}{dx} \int_0^\delta U_L dy = \rho_L U_{Li} \frac{d\delta}{dx} + \int_0^\delta \frac{\partial U_L}{\partial x} dy = \rho_L U_{Li} \frac{d\delta}{dx} - \rho_L V_{Li} \,. \qquad \text{(2-32)}$$

Similarly, that of the vapor is

$$\rho_V \frac{d}{dx} \int_\delta^h U_V dy = -\rho_V U_{Vi} \frac{d\delta}{dx} - \rho_V \{ V_V(h) - V_{Vi} \} . \qquad \text{(2-33)}$$

The sum of mass flowing out through the control surface must be zero, i.e.,

$$\rho_L \frac{d}{dx}\left(\int_0^\delta U_L dy \right) dx + \rho_V \frac{d}{dx}\left(\int_\delta^h U_V dy \right) dx + \rho_V V_V(h) dx = 0 \,. \qquad \text{(2-34)}$$

Substituting Eqs. (2-32) and (2-33) into Eq. (2-34), we can obtain Eq. (2-22).

The increase of the flow rate of the condensate in the section dx is caused by the condensation rate $\dot{m}_x dx = (\dot{m}_{1x} + \dot{m}_{2x})dx$, i.e.,

$$\dot{m}_x = \rho_L \frac{d}{dx} \int_0^\delta U_L dy .\tag{2-35}$$

From this equation and Eq. (2-32) we can obtain Eq. (2-23). Further we can derive Eq. (2-25) from the fact that the sum of the latent heat released and the heat convectively transferred from the vapor boundary layer is transferred to the condensate film at the vapor-liquid interface.

The condensation mass fluxes of component k (=1,2) are evaluated by adding the diffusive mass flow to the convective mass flow at the vapor-liquid interface, i.e.,

$$
\begin{aligned}
\dot{m}_{kx} &= \left\{ \rho_{kV} U_V - \rho_V D \frac{\partial W_{kV}}{\partial x} \right\}_i \frac{d\delta}{dx} - \left\{ \rho_{kV} V_V - \rho_V D \frac{\partial W_{kV}}{\partial y} \right\}_i \\
&\approx \left\{ \rho_{kV} \left(U_V \frac{d\delta}{dx} - V_V \right) + \rho_V D \frac{\partial W_{kV}}{\partial y} \right\}_i, \quad (k = 1, 2) .
\end{aligned}\tag{2-36}
$$

From these equations we can derive Eq. (2-27).

REFERENCES

1. Bird, R. B. , W. E. Stewart, and E. N. Lightfoot, "Transport Phenomena," p.554, John Wiley, New York (1960).

2. Koh, J. C. Y., Laminar Film Condensation of Condensible Gases and Gaseous Mixtures on a Flat Plate, *Proc. 4th U.S. National Congress of Applied Mechanics*, 2, 1327–1336 (1962).

3. Minkowycz, W. J. and E. M. Sparrow, Condensation Heat Transfer in the Presence of Noncondensables, Interfacial Resistance, Superheating, Variable Properties, and Diffusion, *Int. J. Heat Mass Transfer*, 9, 1125–1144 (1966).

3

Similarity Solution of Forced-Convection Condensation of Binary Vapors

The basic equations for forced-convection condensation of a binary vapor mixture correspond to those in which the buoyancy terms, i.e., the second term on the right-hand side of Eqs. (2-2) and (2-5) are neglected. The subscript F for various physical quantities denotes forced-convection condensation. Relevant literature is mentioned in Chapter 6.

3.1 Similarity Transformation

When the stream functions for the condensate film and the vapor boundary layer which are defined, respectively, by

$$U_L = \frac{\partial \Psi_L}{\partial y}, \quad V_L = -\frac{\partial \Psi_L}{\partial x}, \tag{3.1-1a,b}$$

$$U_V = \frac{\partial \Psi_V}{\partial y}, \quad V_V = -\frac{\partial \Psi_V}{\partial x} \tag{3.1-2a,b}$$

are introduced, Eqs. (2-1) and (2-4) are automatically satisfied. Further, we introduce the independent variables η_{FL} and η_{FV}, the dimensionless stream functions $F_{FL}(\eta_{FL})$ and $F_{FV}(\eta_{FV})$, the dimensionless temperatures $\Theta_{FL}(\eta_{FL})$ and $\Theta_{FV}(\eta_{FV})$, and the normalized concentration $\Phi_F(\eta_{FV})$ as

$$\eta_{FL} = y \left(\frac{U_{V\infty}}{\nu_L x} \right)^{\frac{1}{2}}, \tag{3.1-3}$$

$$\eta_{FV} = (y - \delta) \left(\frac{U_{V\infty}}{\nu_V x} \right)^{\frac{1}{2}}, \tag{3.1-4}$$

$$F_{FL}(\eta_{FL}) = \frac{\Psi_L(x,y)}{(\nu_L U_{V\infty} x)^{\frac{1}{2}}}, \tag{3.1-5}$$

$$F_{FV}(\eta_{FV}) = \frac{\Psi_V(x,y)}{(\nu_V U_{V\infty} x)^{\frac{1}{2}}}, \tag{3.1-6}$$

$$\Theta_{FL}(\eta_{FL}) = \frac{T_i - T_L}{T_i - T_w}, \tag{3.1-7}$$

$$\Theta_{FV}(\eta_{FV}) = \frac{T_{V\infty} - T_V}{T_{V\infty} - T_i}, \tag{3.1-8}$$

$$\Phi_F(\eta_{FV}) = \frac{W_{1V} - W_{1V\infty}}{W_{1Vi} - W_{1V\infty}}. \tag{3.1-9}$$

The transformation of Eqs. (2-2), (2-3), and (2-5)–(2-7) by using Eqs. (3.1-1)–(3.1-9) yields the following system of ordinary differential equations:

$$F'''_{FL} + \frac{1}{2} F_{FL} F''_{FL} = 0, \tag{3.1-10}$$

$$\Theta''_{FL} + \frac{1}{2} Pr_L F_{FL} \Theta'_{FL} = 0, \tag{3.1-11}$$

$$F'''_{FV} + \frac{1}{2} F_{FV} F''_{FV} = 0, \tag{3.1-12}$$

$$\Theta''_{FV} + \frac{1}{2} Pr_V F_{FV} \Theta'_{FV} + \frac{Pr_V c_{p12}}{Sc} (W_{1Vi} - W_{1V\infty}) \Theta'_{FV} \Phi'_F = 0, \tag{3.1-13}$$

$$\Phi''_F + \frac{1}{2} Sc F_{FV} \Phi'_F = 0, \tag{3.1-14}$$

where the prime denotes the differential derivative with respect to η_{FL} or η_{FV}, and Pr $(= \nu/\kappa)$ and Sc $(= \nu_V/D)$ are the Prandtl number and Schmidt number, respectively.

The boundary conditions (2-14)–(2-19) are transformed as

at $\eta_{FL} = 0$:

$$F'_{FLw} = 0, \tag{3.1-15}$$

$$F_{FLw} = 0, \tag{3.1-16}$$

$$\Theta_{FLw} = 1; \tag{3.1-17}$$

as $\eta_{FV} \to \infty$:

$$F'_{FV\infty} = 1, \tag{3.1-18}$$

$$\Theta_{FV\infty} = 0, \tag{3.1-19}$$

$$\Phi_{F\infty} = 0. \tag{3.1-20}$$

The condensate film thickness δ corresponds to η_{FLi}, which must be a constant for enabling the similarity transformation. Consequently, the compatibility conditions (2-20)–(2-27) are transformed as

at $\eta_{FL} = \eta_{FLi}$, or $\eta_{FV} = 0$:

$$F'_{FVi} = F'_{FLi}, \tag{3.1-21}$$

$$F''_{FVi} = R F''_{FLi}, \tag{3.1-22}$$

$$F_{FVi} = R F_{FLi} \tag{3.1-23}$$

$$= 2R\left(\dot{M}_{1FL} + \dot{M}_{2FL}\right) = 2R\dot{M}_{FL}, \tag{3.1-24}$$

$$\Theta_{FLi} = 0, \tag{3.1-25}$$

$$\Theta_{FVi} = 1, \tag{3.1-26}$$

$$-\Theta'_{FLi} = \frac{Pr_L \dot{M}_{FL}}{H_i} + \frac{\lambda_V}{\lambda_L}\left(\frac{\nu_L}{\nu_V}\right)^{\frac{1}{2}}\frac{T_{V\infty} - T_i}{T_i - T_w}(-\Theta'_{FVi}), \tag{3.1-27}$$

$$\Phi_{Fi} = 1, \tag{3.1-28}$$

$$-\Phi'_{Fi} = \frac{W_{1Vi}\dot{M}_{2FL} - (1 - W_{1Vi})\dot{M}_{1FL}}{W_{1Vi} - W_{1V\infty}}RSc, \tag{3.1-29}$$

where the $\rho\mu$ ratio R, the dimensionless condensation mass flux $\dot{M}_{kFL}$ for component k ($= 1, 2$), and the dimensionless total condensation mass flux $\dot{M}_{FL}$ are expressed as

$$R = \left(\frac{\rho_L \mu_L}{\rho_V \mu_V}\right)^{\frac{1}{2}}, \tag{3.1-30}$$

$$\dot{M}_{kFL} = \frac{\dot{m}_{kx}x}{\mu_L Re_{Lx}^{\frac{1}{2}}} \quad (k = 1, 2), \tag{3.1-31}$$

$$\dot{M}_{FL} = \frac{\dot{m}_x x}{\mu_L Re_{Lx}^{\frac{1}{2}}}, \tag{3.1-32}$$

$$H_i = \frac{c_{pL}(T_i - T_w)}{\Delta h_V}, \tag{3.1-33}$$

where Re_{Lx} is the two-phase Reynolds number which is defined by

$$Re_{Lx} = \frac{U_{V\infty}x}{\nu_L}. \tag{3.1-34}$$

Equation (2-28) is transformed as

$$W_{kL} = \frac{\dot{M}_{kFL}}{\dot{M}_{FL}} = \frac{2\dot{M}_{kFL}}{F_{FLi}} \quad (k = 1, 2). \tag{3.1-35}$$

The transformation of Eq. (3.1-29) by using Eqs. (3.1-35) and (3.1-23) yields

$$1 - \frac{ScF_{FVi}}{2(-\Phi'_{Fi})} = \frac{1}{W_R}, \tag{3.1-36}$$

where

$$W_R = \frac{W_{1Vi} - W_{1L}}{W_{1V\infty} - W_{1L}}. \tag{3.1-37}$$

In later calculations the use of Eq. (3.1-36) is more convenient than the use of Eq. (3.1-29) (see Section 3.6).

Such a solution [that every physical quantity at a point (x,y) can be expressed by a function of only η_{FL} or η_{FV}] is named a similarity solution. For example, when the temperature distribution in the y direction at an arbitrary x is plotted in the coordinate of Θ versus η_F, we can obtain a single curve in spite of the fact that both the condensate film and the vapor boundary layer develop with x. Therefore, the necessary condition of the existence of such a similarity solution is that x and/or y do not appear explicitly in Eqs. (3.1-10)–(3.1-29). Accordingly, $\dot{M}_{FL}$ and $\dot{M}_{kFL}$ are not a function of x, and $\dot{m}_x$ and $\dot{m}_{kx}$ are proportional to $x^{-\frac{1}{2}}$. Incidentally, the assumptions stated in Chapter 2 are the necessary conditions for the existence of the similarity solution.

The order of the ordinary differential equations (3.1-10)–(3.1-14) is 12, while the number of their boundary conditions is 6 and that of their compatibility conditions is 8 [except Eq. (3.1-24)], i.e., the number of conditions for solving the equations seems to be in excess by 2. However, Eqs. (3.1-25) for Θ_{FL} and (3.1-26) for Θ_{FV} are derived from a continuity condition of temperature (2-24), and Eq. (2-26) is not a given condition but a boundary value which should be obtained. Equation (3.1-28), corresponding to Eq. (2-26), stems from the normalized form of Eq. (3.1-9). Conclusively, in the physical viewpoint, the number of conditions are necessary and sufficient. However, in the mathematical viewpoint, we can solve Eqs. (3.1-10)–(3.1-14) without the conditions of Eqs. (3.1-27) and (3.1-29). In this case T_w and T_i are undetermined (see Section 3.2 and 3.6).

3.2   Procedure of Numerical Calculation

For the given dimensionless numbers R, Pr_L, and Pr_V or Sc (say $c_{p12} = 0$) and for a fixed value of parameter η_{FLi} we can solve Eqs. (3.1-10)–(3.1-14) subject to Eqs. (3.1-15)–(3.1-23), (3.1-25), (3.1-26), and (3.1-28), where T_i and T_w are still unknown. In the practical application of the solution, we have to search for such values of $\dot{M}_{1FL}$, $\dot{M}_{2FL}$, W_{1L}, W_{1Vi}, T_i, and η_{FLi} that Eqs. (3.1-24), (3.1-27), (3.1-29) or (3.1-36), (3.1-35), (2-29), and (2-30) are satisfied for given values of p, $T_{V\infty}$, T_w, and $W_{1V\infty}$, while the process to solve Eqs. (3.1-10)–(3.1-14) is repeated by using corrected values of dimensionless numbers.

A method to solve T_i and T_w together with relevant physical quantities for given values of p, $T_{V\infty}$, and $W_{1V\infty}$ of a given vapor mixture and for a fixed value of η_{FLi} is explained here. The correlation formulas and literature of physical properties used in the calculation are shown in the Appendix. The physical properties in the condensate film are evaluated at the temperature $T_w + (T_i - T_w)/3$ and the concentration W_{1L}, and those in the vapor boundary layer are evaluated at the concentration $(W_{1Vi} + W_{1V\infty})/2$ and the corresponding saturation temperature (see Chapter 9). Although the accuracy of the numerical results is affected by the accuracy of the physical properties used, which will be somewhat corrected in the future, there is no effect on the dimensionless characteristics of the results which are discussed in later sections.

The numerical calculation is performed, e.g., as shown in the flow chart of Fig. 3.2-1.

1. Guessing initial values:

   (1) Designate the values of p, $W_{1V\infty}$, $T_{V\infty}$, and η_{FLi}.

   (2) Guess temperatures $T_w{}^\circ$ and $T_i{}^\circ$.

   (3) Compute W_{1Vi} and W_{1L} by using Eqs. (2-30) and (2-29), respectively.

   (4) Compute the representative physical properties and dimensionless numbers of R, Sc, Pr_V, c_{p12}, and Pr_L corresponding to the above specified state.

2. Solving the momentum equations:

   (5) Guess a boundary value of F''_{FLw} and a calculation domain $\eta_{FV\infty}$.

   (6) Compute the value of F_{FL} from the following equation, which is obtained by solving Eq. (3.1-10) subject to Eqs. (3.1-15) and (3.1-16) by means of the series expansion method,

$$
\begin{aligned}
F_{FL}(\eta_{FL}) \;=\; & \frac{F''_{FLw}}{2}\eta_{FL}^2 - \frac{2}{5!}\left(\frac{F''_{FLw}}{2}\right)^2 \eta_{FL}^5 \\
& + \frac{22}{8!}\left(\frac{F''_{FLw}}{2}\right)^3 \eta_{FL}^8 - \frac{750}{11!}\left(\frac{F''_{FLw}}{2}\right)^4 \eta_{FL}^{11} \\
& + \frac{55794}{14!}\left(\frac{F''_{FLw}}{2}\right)^5 \eta_{FL}^{14} - \frac{7634274}{17!}\left(\frac{F''_{FLw}}{2}\right)^6 \eta_{FL}^{17} \\
& + \cdots .
\end{aligned}
$$

$$\text{(3.2-1)}$$

   (7) Numerically solve Eq. (3.1-12) subject to the boundary values F_{FVi}, F'_{FVi}, and F''_{FVi}, which are obtained from Eqs. (3.2-1) together with Eqs. (3.1-21)–(3.1-23), by means of the Runge-Kutta-Verner method.

   (8) Modify the guessed value F''_{FLw} until the condition (3.1-18) is satisfied within a convergence radius, e.g. , 10^{-5}, by means of the Newton-Raphson method, as Steps (6) to (8) are repeated.

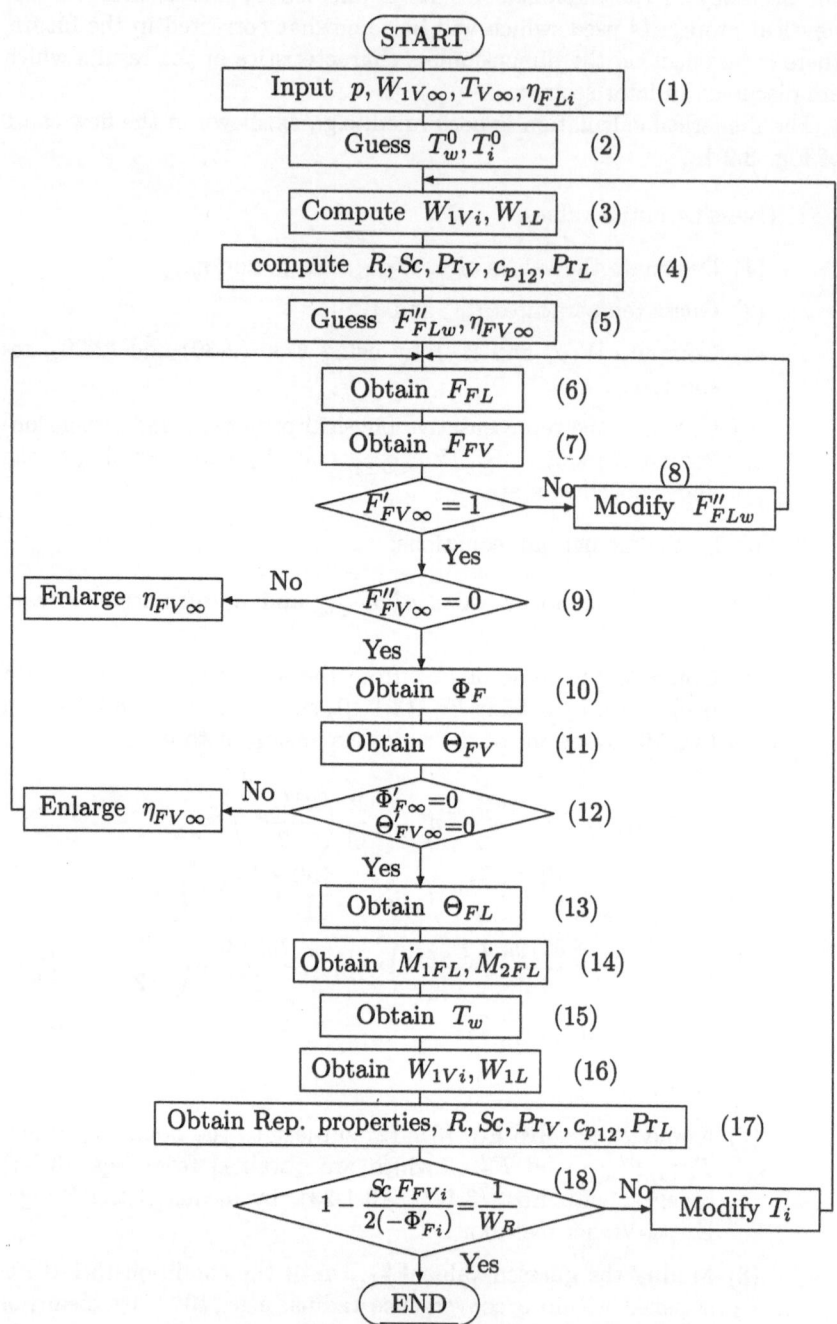

FIGURE 3.2-1. Flow chart for numerically solving the equation system of forced-convection condensation of a binary vapor mixture.

(9) After the value F''_{FLw} is converged, whether the calculation domain $\eta_{FV\infty}$ is sufficient or not is checked by the equation

$$F''_{FL\infty} = 0 \qquad (\text{e.g., } < 10^{-5}).\qquad (3.2\text{-}2)$$

If the guessed value of $\eta_{FV\infty}$ is insufficient, it is enlarged and Steps (6) to (9) are repeated again.

3. Computing Φ_F, Θ_{FV}, Θ_{FL}, and T_w:

(10) Compute Φ_F by substituting the previously obtained F_{FV} into the following equation, which is obtained by integrating Eq. (3.1-14) subject to Eqs. (3.1-20) and (3.1-28),

$$\Phi_F = 1 - \frac{\displaystyle\int_0^{\eta_{FV}} \exp\left(-\frac{Sc}{2}\int_0^{\eta_{FV}} F_{FV} d\eta_{FV}\right) d\eta_{FV}}{\displaystyle\int_0^{\infty} \exp\left(-\frac{Sc}{2}\int_0^{\eta_{FV}} F_{FV} d\eta_{FV}\right) d\eta_{FV}}.$$

$$(3.2\text{-}3)$$

(11) Compute Θ_{FV} by substituting F_{FV} and Φ_F into the following equation, which is obtained by integrating Eq. (3.1-13) subject to Eqs. (3.1-19) and (3.1-26),

$$\Theta_{FV} = 1 - \frac{\displaystyle\int_0^{\eta_{FV}} \exp\left[-\frac{Pr_V}{2}\int_0^{\eta_{FV}}\left\{F_{FV}\right. + \frac{2c_{p12}}{Sc}(W_{1Vi} - W_{1V\infty})\Phi'_F\left.\right\}d\eta_{FV}\right]d\eta_{FV}}{\displaystyle\int_0^{\infty} \exp\left[-\frac{Pr_V}{2}\int_0^{\eta_{FV}}\left\{F_{FV}\right. + \frac{2c_{p12}}{Sc}(W_{1Vi} - W_{1V\infty})\Phi'_F\left.\right\}d\eta_{FV}\right]d\eta_{FV}}.$$

$$(3.2\text{-}4)$$

(12) Check whether the guessed value of $\eta_{FV\infty}$ is sufficient or not for Φ_F and Θ_{FV} by

$$\Phi'_{F\infty} = 0 \qquad\qquad (3.2\text{-}5)$$

and

$$\Theta'_{FV\infty} = 0.\qquad\qquad (3.2\text{-}6)$$

If the value of $\eta_{FV\infty}$ is insufficient, it is enlarged and Steps (6) to (12) are repeated again.

(13) Compute Θ_{FL} by substituting F_{FL} in Eq. (3.2-1) into the following equation, which is derived by integrating Eq. (3.1-11) subject to Eqs. (3.1-17) and (3.1-25),

$$\Theta_{FL} = 1 - \frac{\displaystyle\int_0^{\eta_{FL}} \exp\left(-\frac{\mathrm{Pr}_L}{2}\int_0^{\eta_{FL}} F_{FL}d\eta_{FL}\right)d\eta_{FL}}{\displaystyle\int_0^{\eta_{FLi}} \exp\left(-\frac{\mathrm{Pr}_L}{2}\int_0^{\eta_{FL}} F_{FL}d\eta_{FL}\right)d\eta_{FL}}. \quad (3.2\text{-}7)$$

(14) Compute $\dot{M}_{1FL}$ and $\dot{M}_{2FL}$ by using Eq. (3.1-35), where the prior value of W_{1L} is used in the repetition.

(15) Compute T_w from the following equation, which is derived by rearranging Eq. (3.1-27),

$$T_w = T_i - \frac{1}{-\Theta'_{FLi}}$$

$$\times \left\{ \frac{\mu_L \dot{M}_{FL}\Delta h_V}{\lambda_L} + \frac{\lambda_V}{\lambda_L}\left(\frac{\nu_L}{\nu_V}\right)^{\frac{1}{2}}(T_{V\infty} - T_i)(-\Theta'_{FVi})\right\},$$

$$(3.2\text{-}8)$$

where T_i is the prior value in the repetition.

(16) Compute W_{1Vi} and W_{1L} from Eqs. (2-30) and (2-29).

4. Determining T_i and T_w:

(17) Compute the relevant physical properties and dimensionless numbers R, Sc, Pr_V, c_{p12}, and Pr_L which correspond to the new value of T_w and the prior value T_i in the repetition.

(18) Substitute the obtained values of F_{FVi}, $-\Phi_{Fi}$, W_{1Vi}, and W_{1L} into Eq. (3.1-36). If it is not satisfied, replace the guessed value T_i° to a new value T_i. Then, return to Step (3), and repeat the successive computation. Thus, T_i and T_w are determined together with the other boundary values.

The procedure of calculation for the case where one of the components is noncondensable is similar to the one above and simpler.

3.3   Examples of Numerical Solutions

3.3.1   CHARACTERISTICS OF THE BOUNDARY VALUES

Table 3.3-1 shows the boundary values, relevant dimensionless numbers, heat flux, and condensation mass flux which are numerically solved under specified conditions of p, $T_{V\infty}$, $W_{1V\infty}$, and η_{FLi} for the mixtures of ethanol-water and CFC114-CFC11 and the mixtures of water, ethanol, CFC114, and HCFC22 with air.

From this table we can find some characteristics below.

1. The values of F_{FVi}(proportional to condensation mass flux $\dot{m}_x$) and F''_{FVi}(proportional to the shear stress at the vapor-liquid interface) are slightly affected by R for the same value of η_{FLi}. The values of F''_{FLw} (proportional to the shear stress at the cooling surface) and F'_{FVi}(proportional to the velocity at the vapor-liquid interface) are small as R is large, because the force from vapor main flow acting on the motion of condensate film decreases for large R.

2. The values of $-\Theta'_{FLw}$ and $-\Theta'_{FLi}$ decrease as η_{FLi} increases. The relation of $-\Theta'_{FLw} \approx -\Theta'_{FLi}$ is valid for small η_{FLi}, while $-\Theta'_{FLw} > -\Theta'_{FLi}$ as η_{FLi} increases. This is the effect of the convection term in the energy equation of the condensate film.

3. In the case of saturated vapors the convective heat flux q_{cx} in the vapor phase at the vapor-liquid interface is less than one percent of the heat flux q_{wx} at the cooling surface, while in the case of superheated vapors q_{cx} is appreciable in comparison to q_{wx}.

4. The effect of the diffusion term, the third term on the left hand-side of Eq. (3.1-13), appears only on the values of $-\Theta'_{FVi}$(proportional to q_{cx}). However, the ratio of q_{cx} to q_{wx} is negligibly small except for an air-water mixture (see Section 3.3.3).

Table 3.3-2 shows the boundary values of the solutions of the equations, in which the conditions (3.1-27) and (3.1-29) and the diffusion term in Eq. (3.1-13) are neglected, in the case of imaginary binary vapor mixtures in which R and $Sc = Pr_V$ are arbitrarily given. The method of solution for this case is much easier than the previous case shown in Fig. 3.2-1. This table is useful for determining the functional form of the effect of a dimensionless number upon condensation characteristics.

3.3.2   DISTRIBUTIONS OF VELOCITY, TEMPERATURE, AND CONCENTRATION

Figures 3.3-1(a)–(d) show the distributions of F'_{FL}, F'_{FV}(solid lines), $\Theta_F = (T - T_w)/(T_{V\infty} - T_w)$ (dot-dash line), and Φ_F(broken line), where

TABLE 3.3-1. Boundary values, dimensionless numbers, heat flux, and condensation mass flux for forced-convection condensation of binary vapor mixtures. Nos. (1-1)–(1-13) ethanol-water; Nos. (2-1)–(2-5) CFC114-CFC11; Nos. (3-1)–(3-9) air-water; Nos. (4-1),(4-2) air-ethanol; Nos. (5-1)–(5-4) air-CFC114; and Nos. (6-1),(6-2) air-HCFC22. The subscript "cal" denotes the results from the algebraic solution.

No.	$\dfrac{p}{MPa}$	$\dfrac{T_{s\infty}}{°C}$	$W_{1V\infty}$	η_{FLi}	$\dfrac{\Delta T_{V\infty}}{°C}$	$\dfrac{T_i}{°C}$	$\dfrac{T_w}{°C}$	W_{1Vi}	W_{1L}	R	Sc	Pr_V	Pr_L	$-c_{p12}$
1- 1	0.1	98.0	0.1451	1.50	0	96.37	93.81	0.2615	0.03055	178.0	0.7640	0.9343	1.727	0.1009
1- 2	0.1	98.0	0.1451	1.50	0	96.37	93.81	0.2615	0.03055	178.0	0.7640	0.9343	1.727	0
1- 3	0.1	98.0	0.1451	1.50	100	96.37	93.39	0.2615	0.03055	178.4	0.7640	0.9343	1.735	0.1009
1- 4	0.1	98.0	0.1451	1.97	0	90.15	50.64	0.5514	0.11562	241.9	0.6827	0.9065	4.027	0.1013
1- 5	0.1	98.0	0.1451	1.97	0	90.15	50.66	0.5513	0.11562	241.9	0.6827	0.9065	4.026	0
1- 6	0.1	98.0	0.1451	1.97	100	90.14	44.01	0.5520	0.11595	254.3	0.6826	0.9065	4.455	0.1013
1- 7	0.1	94.0	0.3946	1.00	0	92.32	91.68	0.4704	0.08038	171.0	0.6347	0.8918	2.038	0.1014
1- 8	0.1	94.0	0.3946	1.80	0	84.73	68.50	0.7006	0.26495	220.9	0.5682	0.8741	4.986	0.1013
1- 9	0.1	94.0	0.3946	1.80	0	84.73	68.51	0.7006	0.26495	220.9	0.5682	0.8741	4.986	0
1-10	0.1	94.0	0.3946	1.80	100	84.73	66.08	0.7006	0.26500	225.3	0.5681	0.8741	5.176	0.1013
1-11	0.1	94.0	0.3946	1.91	0	83.52	43.07	0.7278	0.32769	277.5	0.5602	0.8722	8.544	0.1013
1-12	0.1	94.0	0.3946	1.91	0	83.52	43.09	0.7278	0.32769	277.5	0.5602	0.8722	8.541	0
1-13	0.1	94.0	0.3946	1.91	100	83.51	34.41	0.7279	0.32798	300.8	0.5602	0.8722	9.965	0.1013
2- 1	0.5	61.0	0.8178	1.60	0	59.64	46.39	0.8599	0.76706	31.3	0.4331	0.7539	4.957	-0.1977
2- 2	0.5	61.0	0.8178	1.90	0	59.00	16.97	0.8785	0.79486	34.9	0.4314	0.7543	5.308	-0.1973
2- 3	0.5	58.0	0.9059	1.90	0	56.76	14.52	0.9398	0.89266	34.4	0.4162	0.7597	5.508	-0.1942
2- 4	0.7	90.0	0.1809	2.00	0	88.49	9.65	0.2599	0.16674	32.3	0.4826	0.8321	4.238	-0.2479
2- 5	0.7	85.0	0.4353	2.00	0	82.15	1.23	0.5562	0.41337	32.5	0.4767	0.7896	4.768	-0.2240
3- 1	0.1	98.5	0.0625	1.80	0	95.68	88.01	0.1983	0	194.7	0.5213	0.9241	1.934	0.5324
3- 2	0.1	98.5	0.0625	1.90	0	92.56	78.51	0.3222	0	198.3	0.5250	0.9000	2.120	0.5490
3- 3	0.1	96.0	0.1842	1.00	0	95.10	94.55	0.2232	0	183.9	0.5257	0.8956	1.842	0.5522
3- 4	0.1	90.0	0.4068	1.00	0	86.87	86.21	0.4947	0	171.7	0.5417	0.8071	2.034	0.6292
3- 5	0.1	90.0	0.4068	1.00	0	86.87	86.21	0.4947	0	171.7	0.5417	0.8071	2.034	0
3- 6	0.1	90.0	0.4068	1.60	0	63.91	57.66	0.8372	0	188.8	0.5543	0.7547	2.991	0.6931
3- 7	0.1	90.0	0.4068	1.60	0	63.91	57.69	0.8372	0	188.8	0.5543	0.7547	2.990	0
3- 8	0.1	90.0	0.4068	1.60	10	63.91	57.59	0.8372	0	188.9	0.5543	0.7547	2.993	0.6931
3- 9	0.1	90.0	0.4068	1.60	10	63.91	57.63	0.8372	0	188.8	0.5543	0.7547	2.992	0
4- 1	0.05	60.0	0.0409	1.80	0	57.32	34.13	0.1144	0	273.5	0.4154	0.8525	12.860	0.4189
4- 2	0.05	50.0	0.3050	1.20	0	45.02	41.67	0.4230	0	262.4	0.6062	0.7712	12.717	0.4600
5- 1	0.2	18.0	0.0297	1.60	0	15.56	6.64	0.0460	0	65.6	0.2368	0.7177	6.185	0.0000
5- 2	0.5	50.0	0.0201	1.40	0	48.85	41.77	0.0261	0	31.7	0.2062	0.7385	5.402	-0.3014
5- 3	0.5	20.0	0.2295	1.40	0	5.49	-1.55	0.3802	0	57.9	0.6330	0.6294	6.509	-0.3762
5- 4	0.5	20.0	0.2295	1.40	0	5.49	-1.58	0.3802	0	57.9	0.6330	0.6294	6.510	0.0000
6- 1	0.8	12.0	0.0348	1.30	0	11.06	6.71	0.0448	0	24.9	0.2875	0.9091	2.465	-0.2593
6- 2	0.8	12.0	0.0348	1.50	0	10.45	2.48	0.0511	0	25.5	0.2900	0.9029	2.492	-0.2614

TABLE 3.3-1.(2)

No.	$\dfrac{F''_{FLw}}{10^{-2}}$	F_{FVi}	$\dfrac{F'_{FVi}}{10^{-2}}$	F''_{FVi}	$-\Phi'_{Fi}$	$-\Theta'_{FVi}$	$-\Theta'_{FLw}$	$-\Theta'_{FLi}$	$C_F(Sc)$	$C_F(Pr_V)$
1- 1	0.3292	0.6590	0.4936	0.5853	0.4996	0.5678	0.6669	0.6659	0.3010	0.3232
1- 2	0.3292	0.6590	0.4936	0.5853	0.4996	0.5640	0.6669	0.6659	0.3010	0.3232
1- 3	0.3285	0.6591	0.4926	0.5854	0.4996	0.5679	0.6669	0.6659	0.3010	0.3232
1- 4	0.9549	4.4797	1.8783	2.2960	1.6403	2.1854	0.5107	0.4984	0.2892	0.3198
1- 5	0.9549	4.4793	1.8784	2.2958	1.6402	2.1341	0.5107	0.4984	0.2892	0.3198
1- 6	0.9152	4.5140	1.8003	2.3141	1.6510	2.2006	0.5109	0.4978	0.2892	0.3197
1- 7	0.2382	0.2037	0.2382	0.4073	0.3329	0.3894	1.0001	0.9997	0.2818	0.3179
1- 8	0.4433	1.5863	0.7976	0.9773	0.6416	0.9042	0.5570	0.5511	0.2708	0.3157
1- 9	0.4433	1.5863	0.7976	0.9773	0.6416	0.8864	0.5570	0.5511	0.2708	0.3157
1-10	0.4347	1.5867	0.7821	0.9776	0.6417	0.9043	0.5571	0.5510	0.2708	0.3157
1-11	0.5378	2.7220	1.0265	1.4881	0.9157	1.3621	0.5270	0.5132	0.2694	0.3154
1-12	0.5379	2.7219	1.0266	1.4881	0.9157	1.3334	0.5270	0.5132	0.2694	0.3154
1-13	0.4976	2.7296	0.9497	1.4925	0.9175	1.3651	0.5273	0.5124	0.2694	0.3154
2- 1	2.0661	0.8274	3.2999	0.6423	0.3949	0.5497	0.6305	0.6088	0.2455	0.2996
2- 2	3.3863	2.1306	6.4030	1.1600	0.6330	0.9728	0.5398	0.4871	0.2451	0.2996
2- 3	3.4258	2.1250	6.4774	1.1567	0.6153	0.9801	0.5404	0.4852	0.2420	0.3004
2- 4	5.1966	3.3492	10.3042	1.6233	0.9532	1.5130	0.5182	0.4476	0.2553	0.3102
2- 5	5.1786	3.3555	10.2688	1.6265	0.9446	1.4393	0.5204	0.4416	0.2542	0.3045
3- 1	0.5019	1.5832	0.9029	0.9751	0.6025	0.9717	0.5562	0.5536	0.2625	0.3219
3- 2	0.7049	2.5226	1.3380	1.3925	0.8215	1.4135	0.5274	0.5230	0.2632	0.3189
3- 3	0.2216	0.2037	0.2216	0.4073	0.3065	0.3943	1.0001	0.9997	0.2633	0.3184
3- 4	0.2373	0.2037	0.2373	0.4073	0.3105	0.3842	1.0001	0.9997	0.2662	0.3069
3- 5	0.2373	0.2037	0.2373	0.4073	0.3105	0.3704	1.0001	0.9997	0.2662	0.3069
3- 6	0.3518	0.8502	0.5628	0.6635	0.4583	0.6619	0.6256	0.6233	0.2684	0.2997
3- 7	0.3519	0.8502	0.5629	0.6635	0.4583	0.5552	0.6256	0.6233	0.2684	0.2997
3- 8	0.3517	0.8502	0.5625	0.6635	0.4583	0.6619	0.6256	0.6233	0.2684	0.2997
3- 9	0.3518	0.8501	0.5627	0.6634	0.4583	0.5552	0.6256	0.6233	0.2684	0.2997
4- 1	0.3592	1.5913	0.6464	0.9807	0.5146	0.8914	0.5587	0.5463	0.2418	0.3129
4- 2	0.1733	0.3274	0.2079	0.4546	0.3556	0.4128	0.8340	0.8313	0.2772	0.3020
5- 1	1.0026	0.8413	1.6028	0.6553	0.2810	0.5371	0.6283	0.6151	0.1963	0.2944
5- 2	1.6526	0.5132	2.3115	0.5219	0.2331	0.4492	0.7179	0.7034	0.1862	0.2974
5- 3	0.9111	0.5166	1.2748	0.5261	0.4127	0.3999	0.7167	0.7071	0.2815	0.2809
5- 4	0.9109	0.5166	1.2746	0.5261	0.4127	0.4114	0.7167	0.7071	0.2815	0.2809
6- 1	1.9353	0.4076	2.5136	0.4809	0.2626	0.4672	0.7709	0.7642	0.2111	0.3201
6- 2	2.2464	0.6449	3.3643	0.5700	0.2926	0.5516	0.6693	0.6588	0.2118	0.3193

TABLE 3.3-1. (3)

No.	$q_{wx}(\frac{x}{U_{V\infty}})^{1/2}$ 10^3 (Wm^{-2}s$^{-1/2}$)	$\{q_{wx}(\frac{x}{U_{V\infty}})^{1/2}\}_{cal}$ 10^3 (Wm^{-2}s$^{-1/2}$)	$q_{cx}(\frac{x}{U_{V\infty}})^{1/2}$ 10^3 (Wm^{-2}s$^{-1/2}$)	$\{q_{cx}(\frac{x}{U_{V\infty}})^{1/2}\}_{cal}$ 10^3 (Wm^{-2}s$^{-1/2}$)	$\dot{m}_x(\frac{x}{U_{V\infty}})^{1/2}$ 10^{-3} (kgm^{-2}s$^{-1/2}$)	$\{\dot{m}_x(\frac{x}{U_{V\infty}})^{1/2}\}_{cal}$ 10^{-3} (kgm^{-2}s$^{-1/2}$)
1- 1	2.0791	2.0798	0.0055	0.0056	0.9324	0.9312
1- 2	2.0791	2.0797	0.0055	0.0056	0.9324	0.9312
1- 3	2.4198	2.4233	0.3452	0.3506	0.9326	0.9295
1- 4	14.521	14.483	0.109	0.108	6.6554	6.5813
1- 5	14.517	14.479	0.106	0.106	6.6547	6.5808
1- 6	16.086	16.047	1.507	1.493	6.7071	6.6089
1- 7	0.67807	0.67860	0.0043	0.0043	0.3120	0.3121
1- 8	4.9628	4.9679	0.0578	0.0588	2.5429	2.5306
1- 9	4.9617	4.9667	0.0567	0.0576	2.5429	2.5306
1-10	5.5956	5.6109	0.6819	0.6920	2.5435	2.5236
1-11	8.3004	8.2829	0.0991	0.0995	4.3880	4.3376
1-12	8.2982	8.2806	0.0970	0.0974	4.3880	4.3377
1-13	9.3145	9.3027	1.0467	1.0464	4.4003	4.3265
2- 1	1.1179	1.1169	0.0157	0.0156	8.4481	8.3495
2- 2	3.0593	3.0521	0.0410	0.0407	21.8067	21.4350
2- 3	3.0093	2.9991	0.0260	0.0258	22.1973	21.8097
2- 4	6.4736	6.4726	0.0461	0.0457	37.759	37.252
2- 5	6.2452	6.2197	0.0931	0.0920	38.617	37.913
3- 1	5.0879	5.0919	0.0157	0.0160	2.2273	2.2218
3- 2	8.4193	8.4146	0.0491	0.0498	3.6481	3.6285
3- 3	0.6739	0.6739	0.0021	0.0021	0.2961	0.2960
3- 4	0.7711	0.7716	0.0075	0.0075	0.3334	0.3334
3- 5	0.7708	0.7713	0.0073	0.0073	0.3334	0.3334
3- 6	3.7064	3.7181	0.1125	0.1139	1.5254	1.5254
3- 7	3.6882	3.6996	0.0944	0.0956	1.5254	1.5254
3- 8	3.7497	3.7623	0.1556	0.1576	1.5254	1.5253
3- 9	3.7243	3.7365	0.1305	0.1323	1.5253	1.5253
4- 1	2.0659	2.0629	0.0132	0.0133	2.2853	2.2629
4- 2	0.4475	0.4479	0.0113	0.0113	0.4858	0.4850
5- 1	0.7153	0.7128	0.0162	0.0164	5.1744	5.1103
5- 2	0.6474	0.6447	0.0112	0.0112	5.2503	5.1848
5- 3	0.6365	0.6414	0.1090	0.1078	3.8291	3.8278
5- 4	0.6397	0.6448	0.1121	0.1111	3.8291	3.8273
6- 1	0.8081	0.8075	0.0078	0.0077	4.0773	4.0549
6- 2	1.2904	1.2918	0.0152	0.0151	6.4334	6.3965

TABLE 3.3-2. Boundary values for forced-convection condensation of imaginary binary vapor mixtures.

R	$\begin{array}{c}Sc\\=Pr_V\end{array}$	η_{FLi}	$\dfrac{F''_{FLw}}{10^{-2}}$	F_{FVi}	$\dfrac{F'_{FVi}}{10^{-2}}$	F''_{FVi}	$\begin{array}{c}-\Phi'_{Fi}\\=-\Theta'_{FVi}\end{array}$	W_R	$\begin{array}{c}-\Theta'_{FLw}\ -\Theta'_{FLi}\\ \text{for } Pr_L=1\end{array}$	
20	0.2	0.50	1.7385	0.0435	0.8692	0.3476	0.1888	1.0236	2.0001	1.9997
20	0.2	1.80	4.3001	1.3903	7.6999	0.8422	0.3049	1.8380	0.5585	0.5469
20	0.2	2.00	7.0070	2.7898	13.8526	1.3376	0.4231	2.9356	0.5058	0.4828
20	0.2	2.10	9.6232	4.2128	19.8421	1.7874	0.5463	4.3690	0.4850	0.4504
20	0.2	2.15	11.2356	5.1462	23.6107	2.0483	0.6290	5.4982	0.4759	0.4338
20	0.2	2.20	12.9339	6.1897	27.6666	2.3078	0.7231	6.9450	0.4675	0.4171
20	0.2	2.25	14.6264	7.3045	31.8150	2.5484	0.8252	8.7105	0.4597	0.4005
20	0.2	2.30	16.2514	8.4597	35.9144	2.7600	0.9324	10.7834	0.4525	0.3843
20	0.3	0.50	1.7385	0.0435	0.8692	0.3476	0.2214	1.0303	2.0001	1.9997
20	0.3	1.50	2.8480	0.6403	4.2635	0.5651	0.2982	1.4751	0.6680	0.6627
20	0.3	1.80	4.3001	1.3903	7.6999	0.8422	0.3929	2.1310	0.5585	0.5469
20	0.3	2.00	7.0070	2.7898	13.8526	1.3376	0.5732	3.7045	0.5058	0.4828
20	0.3	2.10	9.6232	4.2128	19.8421	1.7874	0.7628	5.8272	0.4850	0.4504
20	0.3	2.15	11.2356	5.1462	23.6107	2.0483	0.8902	7.5243	0.4759	0.4338
20	0.3	2.20	12.9339	6.1897	27.6666	2.3078	1.0350	9.7151	0.4675	0.4171
20	0.3	2.25	14.6264	7.3045	31.8150	2.5484	1.1918	12.4018	0.4597	0.4005
20	0.5	0.50	1.7385	0.0435	0.8692	0.3476	0.2695	1.0420	2.0001	1.9997
20	0.5	1.50	2.8480	0.6403	4.2635	0.5651	0.3933	1.6863	0.6680	0.6627
20	0.5	1.80	4.3001	1.3903	7.6999	0.8422	0.5516	2.7038	0.5585	0.5469
20	0.5	1.90	5.3294	1.9181	10.0494	1.0339	0.6659	3.5722	0.5303	0.5144
20	0.5	2.00	7.0070	2.7898	13.8526	1.3376	0.8596	5.3007	0.5058	0.4828
20	0.5	2.05	8.1937	3.4234	16.5546	1.5454	1.0035	6.7949	0.4950	0.4668
20	0.5	2.10	9.6232	4.2128	19.8421	1.7874	1.1858	8.9426	0.4850	0.4504
20	0.5	2.13	10.5746	4.7574	22.0569	1.9430	1.3131	10.6107	0.4794	0.4404
20	0.7	0.50	1.7385	0.0435	0.8692	0.3476	0.3063	1.0522	2.0001	1.9997
20	0.7	1.50	2.8480	0.6403	4.2635	0.5651	0.4765	1.8880	0.6680	0.6627
20	0.7	1.70	3.6356	1.0491	6.1576	0.7164	0.5967	2.5997	0.5904	0.5817
20	0.7	1.80	4.3001	1.3903	7.6999	0.8422	0.6995	3.2856	0.5585	0.5469
20	0.7	1.95	6.0645	2.2975	11.7163	1.1684	0.9815	5.5330	0.5176	0.4986
20	0.7	2.00	7.0070	2.7898	13.8526	1.3376	1.1387	7.0172	0.5058	0.4828
20	0.7	2.05	8.1937	3.4234	16.5546	1.5454	1.3442	9.2040	0.4950	0.4668
20	0.7	2.10	9.6232	4.2128	19.8421	1.7874	1.6042	12.3637	0.4850	0.4504
20	1.0	0.50	1.7385	0.0435	0.8692	0.3476	0.3507	1.0661	2.0001	1.9997
20	1.0	1.00	2.0281	0.2028	2.0272	0.4049	0.4133	1.3250	1.0004	0.9987
20	1.0	1.50	2.8480	0.6403	4.2635	0.5651	0.5902	2.1854	0.6680	0.6627
20	1.0	1.70	3.6356	1.0491	6.1576	0.7164	0.7634	3.1964	0.5904	0.5817
20	1.0	1.80	4.3001	1.3903	7.6999	0.8422	0.9125	4.1984	0.5585	0.5469
20	1.0	1.90	5.3294	1.9181	10.0494	1.0339	1.1495	6.0364	0.5303	0.5144
20	1.0	2.00	7.0070	2.7898	13.8526	1.3376	1.5527	9.8410	0.5058	0.4828
20	1.0	2.03	7.6886	3.1516	15.4021	1.4577	1.7231	11.6978	0.4992	0.4732

TABLE 3.3-2.(2)

R	Sc $= Pr_V$	η_{FLi}	$\dfrac{F''_{FLw}}{10^{-2}}$	F_{FVi}	$\dfrac{F'_{FVi}}{10^{-2}}$	F''_{FVi}	$-\Phi'_{Fi}$ $=-\Theta'_{FVi}$	W_R	$-\Theta'_{FLw}$ $-\Theta'_{FLi}$ for $Pr_L=1$	
50	0.2	0.50	0.6956	0.0435	0.3478	0.3478	0.1883	1.0236	2.0000	1.9999
50	0.2	1.70	1.5237	1.1002	2.5863	0.7571	0.2783	1.6537	0.5892	0.5855
50	0.2	1.90	2.4909	2.2448	4.7159	1.2278	0.3740	2.5015	0.5282	0.5207
50	0.2	1.98	3.5303	3.4522	6.9503	1.7253	0.4768	3.6228	0.5079	0.4965
50	0.2	2.01	4.1810	4.2110	8.3449	2.0323	0.5429	4.4584	0.5010	0.4871
50	0.2	2.03	4.7302	4.8572	9.5240	2.2885	0.5999	5.2542	0.4967	0.4806
50	0.2	2.07	6.1193	6.5257	12.5257	2.9246	0.7502	7.6810	0.4885	0.4670
50	0.2	2.10	7.3595	8.0683	15.2394	3.4771	0.8925	10.4173	0.4829	0.4563
50	0.3	0.50	0.6956	0.0435	0.3478	0.3478	0.2207	1.0305	2.0000	1.9999
50	0.3	1.70	1.5237	1.1002	2.5863	0.7571	0.3531	1.8775	0.5892	0.5855
50	0.3	1.90	2.4909	2.2448	4.7159	1.2278	0.4982	3.0852	0.5282	0.5207
50	0.3	1.97	3.3531	3.2463	6.5705	1.6411	0.6289	4.4313	0.5103	0.4995
50	0.3	2.00	3.9424	3.9321	7.8335	1.9201	0.7203	5.5216	0.5033	0.4902
50	0.3	2.03	4.7302	4.8572	9.5240	2.2885	0.8457	7.2193	0.4967	0.4806
50	0.3	2.05	5.3787	5.6293	10.9211	2.5877	0.9521	8.8414	0.4925	0.4739
50	0.3	2.07	6.1193	6.5257	12.5257	2.9246	1.0771	10.9636	0.4885	0.4670
50	0.5	0.50	0.6956	0.0435	0.3478	0.3478	0.2685	1.0422	2.0000	1.9999
50	0.5	1.50	1.1612	0.6530	1.7404	0.5787	0.3921	1.7132	0.6672	0.6650
50	0.5	1.80	1.8643	1.5087	3.3482	0.9238	0.5716	2.9402	0.5568	0.5518
50	0.5	1.90	2.4909	2.2448	4.7159	1.2278	0.7316	4.2930	0.5282	0.5207
50	0.5	1.96	3.1926	3.0600	6.2263	1.5646	0.9139	6.1376	0.5128	0.5026
50	0.5	1.98	3.5303	3.4522	6.9503	1.7253	1.0031	7.1600	0.5079	0.4965
50	0.5	2.01	4.1810	4.2110	8.3449	2.0323	1.1782	9.3943	0.5010	0.4871
50	0.5	2.03	4.7302	4.8572	9.5240	2.2885	1.3292	11.5666	0.4967	0.4806
50	0.7	0.50	0.6956	0.0435	0.3478	0.3478	0.3050	1.0525	2.0000	1.9999
50	0.7	1.50	1.1612	0.6530	1.7404	0.5787	0.4752	1.9266	0.6672	0.6650
50	0.7	1.70	1.5237	1.1002	2.5863	0.7571	0.6057	2.7453	0.5892	0.5855
50	0.7	1.80	1.8643	1.5087	3.3482	0.9238	0.7285	3.6341	0.5568	0.5518
50	0.7	1.90	2.4909	2.2448	4.7159	1.2278	0.9567	5.5941	0.5282	0.5207
50	0.7	1.94	2.9147	2.7376	5.6296	1.4318	1.1133	7.1745	0.5177	0.5087
50	0.7	1.98	3.5303	3.4522	6.9503	1.7253	1.3446	9.8593	0.5079	0.4965
50	0.7	2.00	3.9424	3.9321	7.8335	1.9201	1.5022	11.9275	0.5033	0.4902
50	1.0	0.50	0.6956	0.0435	0.3478	0.3478	0.3490	1.0664	2.0000	1.9999
50	1.0	1.50	1.1612	0.6530	1.7404	0.5787	0.5890	2.2438	0.6672	0.6650
50	1.0	1.70	1.5237	1.1002	2.5863	0.7571	0.7772	3.4220	0.5892	0.5855
50	1.0	1.80	1.8643	1.5087	3.3482	0.9238	0.9558	4.7458	0.5568	0.5518
50	1.0	1.85	2.1233	1.8147	3.9172	1.0498	1.0926	5.8974	0.5421	0.5360
50	1.0	1.90	2.4909	2.2448	4.7159	1.2278	1.2886	7.7530	0.5282	0.5207
50	1.0	1.93	2.7942	2.5977	5.3704	1.3739	1.4519	9.4853	0.5203	0.5117
50	1.0	1.94	2.9147	2.7376	5.6296	1.4318	1.5172	10.2246	0.5177	0.5087

TABLE 3.3-2. (3)

R	Sc $= Pr_V$	η_{FLi}	$\dfrac{F''_{FLw}}{10^{-2}}$	F_{FVi}	$\dfrac{F'_{FVi}}{10^{-2}}$	F''_{FVi}	$-\Phi'_{Fi}$ $=-\Theta'_{FVi}$	W_R	$-\Theta'_{FLw}$ for	$-\Theta'_{FLi}$ $Pr_L=1$
100	0 2	0.50	0.3478	0.0435	0.1739	0.3478	0.1881	1.0237	2.0000	1.9999
100	0.2	1.70	0.7746	1.1190	1.3158	0.7722	0.2791	1.6691	0.5887	0.5868
100	0.2	1.90	1.3388	2.4147	2.5389	1.3286	0.3873	2.6560	0.5273	0.5233
100	0.2	1.95	1.7296	3.2848	3.3637	1.7112	0.4613	3.4731	0.5142	0.5087
100	0.2	2.00	2.5267	5.0449	5.0323	2.4845	0.6154	5.5504	0.5021	0.4937
100	0.2	2.02	3.0746	6.2597	6.1782	3.0105	0.7247	7.3385	0.4977	0.4873
100	0.2	2.03	3.4154	7.0205	6.8922	3.3351	0.7943	8.6126	0.4955	0.4839
100	0.2	2.04	3.8005	7.8870	7.7014	3.6998	0.8743	10.2104	0.4935	0.4804
100	0.3	0.50	0.3478	0.0435	0.1739	0.3478	0.2204	1.0305	2.0000	1.9999
100	0.3	1.50	0.5843	0.6572	0.8761	0.5833	0.2970	1.4969	0.6669	0.6658
100	0.3	1.80	0.9616	1.5571	1.7289	0.9571	0.4093	2.3292	0.5562	0.5536
100	0.3	1.91	1.3995	2.5507	2.6677	1.3882	0.5363	3.4902	0.5246	0.5204
100	0.3	1.97	1.9759	3.8293	3.8803	1.9512	0.7048	5.4048	0.5092	0.5028
100	0.3	2.00	2.5267	5.0449	5.0323	2.4845	0.8697	7.6999	0.5021	0.4937
100	0.3	2.01	2.7792	5.6037	5.5602	2.7275	0.9467	8.9169	0.4998	0.4905
100	0.3	2.02	3.0746	6.2597	6.1782	3.0105	1.0380	10.4792	0.4977	0.4873
100	0.5	0.50	0.3478	0.0435	0.1739	0.3478	0.2682	1.0422	2.0000	1.9999
100	0.5	1.50	0.5843	0.6572	0.8761	0.5833	0.3917	1.7226	0.6669	0.6658
100	0.5	1.70	0.7746	1.1190	1.3158	0.7722	0.4873	2.3481	0.5887	0.5868
100	0.5	1.85	1.1113	1.9006	2.0529	1.1048	0.6541	3.6548	0.5413	0.5382
100	0.5	1.92	1.4674	2.7024	2.8114	1.4548	0.8310	5.3463	0.5220	0.5175
100	0.5	1.95	1.7296	3.2848	3.3637	1.7112	0.9625	6.8120	0.5142	0.5087
100	0.5	1.97	1.9759	3.8293	3.8803	1.9512	1.0872	8.3696	0.5092	0.5028
100	0.5	2.00	2.5267	5.0449	5.0323	2.4845	1.3707	12.5203	0.5021	0.4937
100	0.7	0.50	0.3478	0.0435	0.1739	0.3478	0.3046	1.0526	2.0000	1.9999
100	0.7	1.50	0.5843	0.6572	0.8761	0.5833	0.4747	1.9401	0.6669	0.6658
100	0.7	1.70	0.7746	1.1190	1.3158	0.7722	0.6091	2.8009	0.5887	0.5868
100	0.7	1.80	0.9616	1.5571	1.7289	0.9571	0.7408	3.7835	0.5562	0.5536
100	0.7	1.85	1.1113	1.9006	2.0529	1.1048	0.8464	4.6726	0.5413	0.5382
100	0.7	1.91	1.3995	2.5507	2.6677	1.3882	1.0507	6.6534	0.5246	0.5204
100	0.7	1.94	1.6305	3.0653	3.1554	1.6145	1.2158	8.5076	0.5167	0.5116
100	0.7	1.96	1.8436	3.5371	3.6030	1.8224	1.3692	10.4356	0.5117	0.5058
100	1.0	0.50	0.3478	0.0435	0.1739	0.3478	0.3484	1.0665	2.0000	1.9999
100	1.0	1.00	0.4072	0.2036	0.4072	0.4071	0.4087	1.3317	1.0001	0.9997
100	1.0	1.50	0.5843	0.6572	0.8761	0.5833	0.5885	2.2645	0.6669	0.6658
100	1.0	1.70	0.7746	1.1190	1.3158	0.7722	0.7825	3.5092	0.5887	0.5868
100	1.0	1.80	0.9616	1.5571	1.7289	0.9571	0.9740	4.9838	0.5562	0.5536
100	1.0	1.85	1.1113	1.9006	2.0529	1.1048	1.1280	6.3488	0.5413	0.5382
100	1.0	1.90	1.3388	2.4147	2.5389	1.3286	1.3632	8.7456	0.5273	0.5233
100	1.0	1.92	1.4674	2.7024	2.8114	1.4548	1.4969	10.2756	0.5220	0.5175

TABLE 3.3-2. (4)

R	$\frac{Sc}{=Pr_V}$	η_{FLi}	$\frac{F''_{FLw}}{10^{-2}}$	F_{FVi}	$\frac{F'_{FVi}}{10^{-2}}$	F''_{FVi}	$\begin{array}{c}-\Phi'_{Fi}\\=-\Theta'_{FVi}\end{array}$	W_R	$-\Theta'_{FLw}$ $-\Theta'_{FLi}$ for $Pr_L=1$	
500	0.2	0.50	0.0696	0.0435	0.0348	0.3479	0.1879	1.0237	2.0000	2.0000
500	0.2	1.00	0.0815	0.2037	0.0815	0.4075	0.2018	1.1123	1.0000	0.9999
500	0.2	1.70	0.1571	1.1347	0.2670	0.7848	0.2798	1.6822	0.5883	0.5880
500	0.2	1.84	0.2237	1.8928	0.4114	1.1170	0.3427	2.2334	0.5436	0.5430
500	0.2	1.90	0.2879	2.5977	0.5468	1.4371	0.4019	2.8279	0.5265	0.5257
500	0.2	1.94	0.3719	3.4981	0.7210	1.8551	0.4787	3.7132	0.5158	0.5146
500	0.2	1.97	0.5029	4.8775	0.9899	2.5064	0.5995	5.3663	0.5080	0.5064
500	0.2	1.99	0.6968	6.8954	1.3851	3.4681	0.7817	8.4827	0.5031	0.5008
500	0.2	2.00	0.8828	8.8226	1.7630	4.3880	0.9605	12.2759	0.5007	0.4978
500	0.3	0.50	0.0696	0.0435	0.0348	0.3479	0.2202	1.0305	2.0000	2.0000
500	0.3	1.70	0.1571	1.1347	0.2670	0.7848	0.3554	1.9189	0.5883	0.5880
500	0.3	1.80	0.1976	1.6003	0.3556	0.9870	0.4137	2.3822	0.5557	0.5552
500	0.3	1.84	0.2237	1.8928	0.4114	1.1170	0.4507	2.7025	0.5436	0.5430
500	0.3	1.90	0.2879	2.5977	0.5468	1.4371	0.5411	3.5726	0.5265	0.5257
500	0.3	1.95	0.4050	3.8486	0.7892	2.0197	0.7060	5.4851	0.5131	0.5119
500	0.3	1.97	0.5029	4.8775	0.9899	2.5064	0.8453	7.4372	0.5080	0.5064
500	0.3	1.98	0.5807	5.6897	1.1488	2.8928	0.9571	9.2319	0.5055	0.5036
500	0.3	1.99	0.6968	6.8954	1.3851	3.4681	1.1257	12.3230	0.5031	0.5008
500	0.5	0.50	0.0696	0.0435	0.0348	0.3479	0.2679	1.0423	2.0000	2.0000
500	0.5	1.50	0.1174	0.6606	0.1762	0.5870	0.3913	1.7302	0.6667	0.6665
500	0.5	1.70	0.1571	1.1347	0.2670	0.7848	0.4892	2.3803	0.5883	0.5880
500	0.5	1.82	0.2096	1.7351	0.3813	1.0467	0.6167	3.3716	0.5496	0.5490
500	0.5	1.90	0.2879	2.5977	0.5468	1.4371	0.8058	5.1536	0.5265	0.5257
500	0.5	1.93	0.3451	3.2127	0.6656	1.7218	0.9441	6.6993	0.5184	0.5173
500	0.5	1.95	0.4050	3.8486	0.7892	2.0197	1.0896	8.5510	0.5131	0.5119
500	0.5	1.96	0.4471	4.2925	0.8757	2.2291	1.1924	10.0008	0.5106	0.5091
500	0.7	0.50	0.0696	0.0435	0.0348	0.3479	0.3043	1.0527	2.0000	2.0000
500	0.7	1.50	0.1174	0.6606	0.1762	0.5870	0.4743	1.9511	0.6667	0.6665
500	0.7	1.70	0.1571	1.1347	0.2670	0.7848	0.6120	2.8484	0.5883	0.5880
500	0.7	1.80	0.1976	1.6003	0.3556	0.9870	0.7519	3.9203	0.5557	0.5552
500	0.7	1.86	0.2406	2.0807	0.4474	1.2015	0.9002	5.2367	0.5378	0.5371
500	0.7	1.90	0.2879	2.5977	0.5468	1.4371	1.0632	6.9041	0.5265	0.5257
500	0.7	1.92	0.3228	2.9747	0.6196	1.6111	1.1840	8.2896	0.5211	0.5201
500	0.7	1.94	0.3719	3.4981	0.7210	1.8551	1.3538	10.4573	0.5158	0.5146
500	1.0	0.50	0.0696	0.0435	0.0348	0.3479	0.3480	1.0666	2.0000	2.0000
500	1.0	1.00	0.0815	0.2037	0.0815	0.4075	0.4078	1.3330	1.0000	0.9999
500	1.0	1.50	0.1174	0.6606	0.1762	0.5870	0.5881	2.2814	0.6667	0.6665
500	1.0	1.70	0.1571	1.1347	0.2670	0.7848	0.7869	3.5841	0.5883	0.5880
500	1.0	1.80	0.1976	1.6003	0.3556	0.9870	0.9905	5.2033	0.5557	0.5552
500	1.0	1.84	0.2237	1.8928	0.4114	1.1170	1.1216	6.4019	0.5436	0.5430
500	1.0	1.88	0.2615	2.3098	0.4913	1.3054	1.3118	8.3611	0.5321	0.5313
500	1.0	1.91	0.3040	2.7724	0.5805	1.5175	1.5264	10.8887	0.5238	0.5229

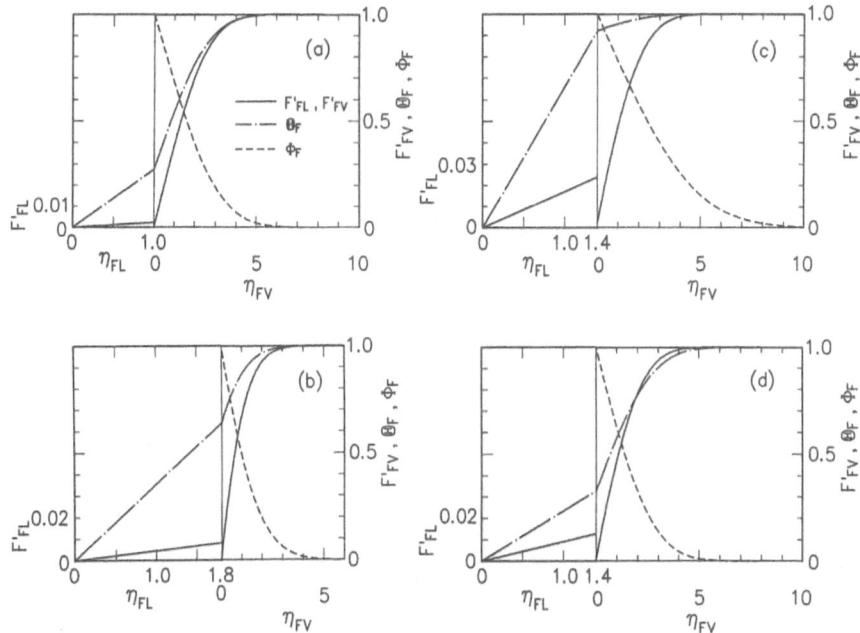

FIGURE 3.3-1. Distributions of velocities F'_{FL}, F'_{FV}, temperature Θ_F, and mass concentration Φ_F across the condensate film and the vapor boundary layer in the case of forced-convection condensation (No. corresponds to that in Table 3.3-1). (a) No. 1-7: ethanol-water, $R = 170$, $Sc = 0.64$, $Pr_V = 0.89$. (b) No. 1-8: ethanol-water, $R=220$, $Sc = 0.57$, $Pr_V = 0.87$. (c) No. 5-2: air-CFC114, $R = 32$, $Sc = 0.21$, $Pr_V = 0.74$. (d) No. 5-3: air-CFC114, $R = 58$, $Sc = 0.63$, $Pr_V = 0.63$.

(a), (b), (c), and (d) correspond to Nos. 1-7, 1-8, 5-2, and 5-3 in Table 3.3-1, respectively. The distributions of F'_{FL} and Θ_F in the condensate film are almost linear [cf. Eq. (3.2-1)]. The vapor boundary layer thickness becomes thinner for a larger value of η_{FLi}, because the suction velocity $(-V_V)$ increases with the increase of the dimensionless condensation mass flux M_{FL}. In the case of a small value of Sc, the thickness of the concentration boundary layer is larger than that of the velocity boundary layer. The above tendencies are valid for the other solution shown in Table 3.3-1. In the case of a large value of η_{FLi}, the values of $(T_i - T_w)$ and $(T_i - T_w)/(T_{V\infty} - T_w)$ increase.

3.3.3   POSSIBILITY OF THE APPEARANCE OF SUBCOOLING IN THE VAPOR BOUNDARY LAYER

Figures 3.3-2(a)–(g) show the distributions of the dimensionless temperature difference $(\Theta_{sFV} - \Theta_{FV}) = (T_V - T_{Vs})/(T_{V\infty} - T_i)$, where the di-

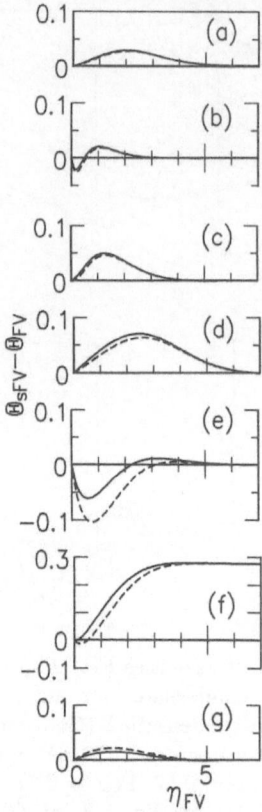

FIGURE 3.3-2. Distributions of $\Theta_{sFV} - \Theta_{FV}$ across the vapor boundary layer in the case of forced-convection condensation (No. corresponds to that in Table 3.3-1). (a) No. 1- 1, 2: ethanol-water, $\eta_{FLi} = 1.5$, $T_{V\infty} = 98°C$, $W_{1V\infty} = 0.145$, $W_{1Vi} = 0.262$. (b) No. 1- 4, 5: ethanol-water, $\eta_{FLi} = 1.97$, $T_{V\infty} = 98°C$, $W_{1V\infty} = 0.145$, $W_{1Vi} = 0.551$. (c) No. 1-11,12: ethanol-water, $\eta_{FLi} = 1.91$, $T_{V\infty} = 94°C$, $W_{1V\infty} = 0.395$, $W_{1Vi} = 0.728$. (d) No. 3- 4, 5: air-water, $\eta_{FLi} = 1.00$, $T_{V\infty} = 90°C$, $W_{1V\infty} = 0.407$, $W_{1Vi} = 0.495$. (e) No. 3- 6, 7: air-water, $\eta_{FLi} = 1.60$, $T_{V\infty} = 90°C$, $W_{1V\infty} = 0.407$, $W_{1Vi} = 0.837$. (f) No. 3- 8, 9: air-water, $\eta_{FLi} = 1.60$, $T_{V\infty} = 90°C$, $W_{1V\infty} = 0.407$, $W_{1Vi} = 0.837$, $\Delta T_{V\infty} = 10$ K. (g) No. 5- 3, 4: air-CFC114, $\eta_{FLi} = 1.40$, $T_{V\infty} = 20°C$, $W_{1V\infty} = 0.230$, $W_{1Vi} = 0.380$.

mensionless saturation temperature $\Theta_{sFV} = (T_{V\infty} - T_{Vs})/(T_{V\infty} - T_i)$ is obtained from the concentration distribution. Solid lines correspond to the solution of Eqs. (3.1-10)– (3.1-14) and broken lines correspond to the case where the diffusion term, i.e., the third term on the left-hand side of Eq. (3.1-13) is neglected. Positive and negative values of $\Theta_{sFV} - \Theta_{FV}$ represent superheated and subcooled states, respectively. It is postulated that the subcooled state is unstable and results in an appearance of mist or fog, and consequently the actual temperature will approach the saturation temperature (see Section 6.1.2).

In Figs. 3.3-2(a), (b), and (c) for ethanol-water mixtures, the difference between solid and broken lines is very small, i.e., the effect of the diffusion term is small. In comparison between (a) and (b), where the common parameters are $T_{V\infty}$ and $W_{1V\infty}$, a subcooled region appears in the vicinity of the vapor-liquid interface in (b), where the value of η_{FLi} is larger. In comparison between (b) and (c), where the parameter η_{FLi} is almost the same, there is no subcooled region in (c), where the value of W_{1Vi} is larger. These differences are related to the range of variation of W_{1V} in the boundary layer and to the gradient of the dew-point line in the phase diagram [see Fig. 6.3-2 (a)].

In Figs. 3.3-2(d), (e) and (f) for air-water mixtures, the difference between solid and broken lines is relatively large. In the comparison between (d) and (e), where the values of $T_{V\infty}$ and $W_{1V\infty}$ are common, a subcooled region appears in (e), where the value of η_{FLi} is larger. However, when the bulk vapor corresponding to that in (e) is superheated by 10 K, the subcooled region disappears as seen in (f). Further, we find in the comparison between the result in (g) for an air-CFC114 mixture and the others that the values of $(\Theta_{sFV} - \Theta_{FV})$ for the solid lines are larger than those for broken lines for a positive value of c_{p12} and it is reversed for a negative value of c_{p12}.

For a prediction of the appearance of subcooling in the vapor boundary layer, we need a more precise theoretical analysis using an accurate phase diagram, considering the diffusion term, and accounting for variation of physical properties with concentration and temperature.

3.4 Formulas of the Boundary Values $\Theta'_{FLw}, \Theta'_{FLi}$, and Θ'_{FVi} for Dimensionless Temperature

The local heat transfer coefficients corresponding to the local values of heat flux at the wall q_{wx}, heat flux at the vapor-liquid interface in the liquid side q_{ix}, and that in the vapor side q_{cx} are defined, respectively, as

$$q_{wx} = \lambda_L \left(\frac{\partial T_L}{\partial y} \right)_w = \alpha_{wx}(T_i - T_w), \qquad (3.4\text{-}1)$$

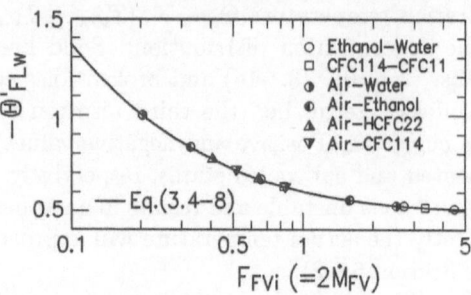

FIGURE 3.4-1. Correlation of $-\Theta'_{FLw}$ versus $F_{FVi}(=2\dot{M}_{FV})$.

$$q_{ix} = \lambda_L \left(\frac{\partial T_L}{\partial y} \right)_i = \alpha_{ix}(T_i - T_w), \tag{3.4-2}$$

$$q_{cx} = \lambda_V \left(\frac{\partial T_V}{\partial y} \right)_i = \alpha_{cx}(T_{V\infty} - T_i). \tag{3.4-3}$$

The dimensionless numbers Nu_{Lwx}, Nu_{Lix}, and Nu_{cx} corresponding to α_{wx}, α_{ix}, and α_{cx} can be expressed as

$$Nu_{Lwx} = \frac{\alpha_{wx}x}{\lambda_L} = (-\Theta'_{FLw})Re_{Lx}^{\frac{1}{2}}, \tag{3.4-4}$$

$$Nu_{Lix} = \frac{\alpha_{ix}x}{\lambda_L} = (-\Theta'_{FLi})Re_{Lx}^{\frac{1}{2}}, \tag{3.4-5}$$

$$Nu_{cx} = \frac{\alpha_{cx}x}{\lambda_V} = (-\Theta'_{FVi})Re_{Vx}^{\frac{1}{2}}, \tag{3.4-6}$$

where

$$Re_{Vx} = \frac{U_{V\infty}x}{\nu_V}. \tag{3.4-7}$$

As for the heat transfer in the condensate film, Fig. 3.4-1 shows the relation of $-\Theta'_{FLw}$ versus $F_{FVi}(= 2\dot{M}_{FV})$ in Table 3.3-1. The solid line in the figure represents the following equation, which was obtained by Fujii et al.:[1]

$$-\Theta'_{FLw} = 0.433 \left(1.367 - \frac{0.432}{\sqrt{2\dot{M}_{FV}}} + \frac{1}{2\dot{M}_{FV}} \right)^{\frac{1}{2}}, \tag{3.4-8}$$

where

$$\dot{M}_{FV} = \frac{1}{2}F_{FVi} = \frac{(\dot{m}_{1x} + \dot{m}_{2x})x}{\mu_V Re_{Vx}^{\frac{1}{2}}} = R\dot{M}_{FL}. \tag{3.4-9}$$

Equation (3.4-8) can correlate the data in maximum error by about two percent.

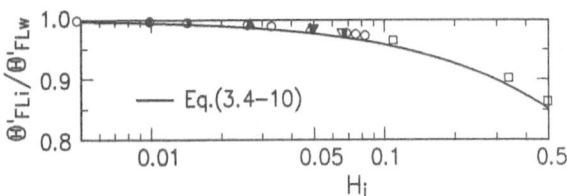

FIGURE 3.4-2. Correlation of $\Theta'_{FLi}/\Theta'_{FLw}$ versus H_i. Symbols used are the same as in Fig. 3.4-1.

In Fig. 3.4-2 the values of $\Theta'_{FLi}/\Theta'_{FLw}$ are plotted against H_i for the data in Table 3.3-1. For the large values of H_i, i.e., for large values of $(T_i - T_w)$, $-\Theta'_{FLi}$ becomes smaller than $-\Theta'_{FLw}$, which means that the effect of convection terms in the condensate film becomes marked. The solid line in the figure represents

$$\frac{\Theta'_{FLi}}{\Theta'_{FLw}} = (1 + 0.320 H_i^{0.87})^{-1} . \tag{3.4-10}$$

This equation correlates the data in maximum error by about two percent (see Section 5.1).

As for the convective heat transfer in the vapor boundary layer, we first consider the case where the condensate film is so thin that T_i is nearly equal to T_w. In this case we can derive the same basic equations and their boundary conditions as in the case of single phase convective heat transfer, i.e., Eqs. (3.1-12) and (3.1-13), in which the diffusion term is neglected, $F_{FVi} = 0$, $F'_{FVi} = 0$, Eqs. (3.1-18), (3.1-26), and (3.1-19). The solution of these equations was correlated by Rose[2] as

$$Nu_{cx} = C_F(\mathrm{Pr}_V)\mathrm{Re}_{Vx}^{\frac{1}{2}} , \tag{3.4-11}$$

where

$$C_F(\mathrm{Pr}_V) = \frac{\mathrm{Pr}_V^{\frac{1}{2}}}{(27.8 + 75.9\mathrm{Pr}_V^{0.306} + 657\mathrm{Pr}_V)^{\frac{1}{6}}} \tag{3.4-12a}$$

$$\approx 0.331\mathrm{Pr}_V^{0.363} \quad (0.2 \lesssim \mathrm{Pr}_V \lesssim 1) \tag{3.4-12b}$$

$$\approx 0.331\mathrm{Pr}_V^{0.344} \quad (0.7 \lesssim \mathrm{Pr}_V \lesssim 10) . \tag{3.4-12c}$$

From Eqs. (3.4-6) and (3.4-11) we can derive the relation

$$-\Theta'_{FVi} = C_F(\mathrm{Pr}_V) . \tag{3.4-13}$$

The value of $-\Theta'_{FVi}$ in condensation will be affected by the condensation mass flux. Figure 3.4-3 shows the results in Table 3.3-2, where the diffusion term in Eq. (3.1-13) is neglected in the coordinates of $\{-\Theta'_{FVi}/C_F(\mathrm{Pr}_V)\}-$

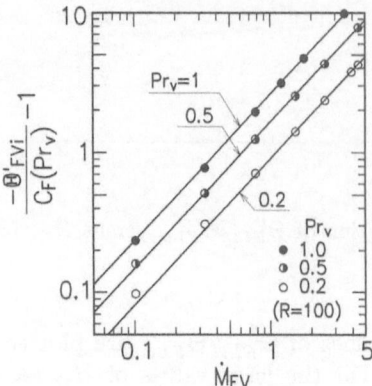

FIGURE 3.4-3. Correlation of $\{-\Theta'_{FVi}/C_F(Pr_V)\} - 1$ versus $\dot{M}_{FV}(=F_{FVi}/2)$.

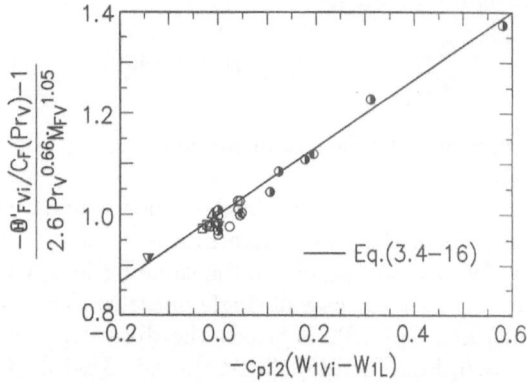

FIGURE 3.4-4.   Correlation of $[\{-\Theta'_{FVi}/C_F(Pr_V)\} - 1]/2.6Pr_V^{0.66}\,\dot{M}_{FV}^{1.05}$ versus $-c_{p12}(W_{1Vi} - W_{1L})$. Symbols used are the same as in Fig. 3.4-1.

1 versus $\dot{M}_{FV}$. The solid line in the figure represents

$$\frac{-\Theta'_{FVi}}{C_F(Pr_V)} = 1 + 2.6Pr_V^{0.66}\dot{M}_{FV}^{1.05}\,. \qquad (3.4\text{-}14)$$

This equation[3] can correlate the data in maximum error by about two percent.

Figure 3.4-4 shows the result in Table 3.3-1, where the diffusion term is taken into account, in the coordinates of $\{-\Theta'_{FVi}/C_F(Pr_V)-1\}/(2.6Pr_V^{0.66}$ $\dot{M}_{FV}^{1.05})$ versus $-c_{p12}(W_{1Vi} - W_{1L})$. The ordinate is derived by referring to Eq. (3.4-14) and the abscissa is taken by referring to the following equation:

$$\Theta''_{FVi} + \frac{1}{2}Pr_V F_{FVi}\{1 - c_{p12}(W_{1Vi} - W_{1L})\}\Theta'_{FVi} = 0\,, \qquad (3.4\text{-}15)$$

which can be derived by combining the value of Eq. (3.1-13) at the vapor-liquid interface with Eq. (3.1-36). The solid line in the figure represents

$$\frac{-\Theta'_{FVi}}{C_F(Pr_V)} = 1 + 2.6Pr_V^{0.66}\dot{M}_{FV}^{1.05}\left\{1 - \frac{2}{3}c_{p12}(W_{1Vi} - W_{1L})\right\}. \quad (3.4\text{-}16)$$

This equation[3] can correlate the data in maximum error by about three percent.

3.5   Formula of the Boundary Value Φ'_{Fi} for Normalized Concentration

Similarly, as in the derivation of the convective heat transfer characteristics in the previous section, we first consider the case of a thin condensate film, where W_{1Vi} corresponds to the value at T_w. The system of the basic equations and their boundary conditions, Eqs. (3.1-12) and (3.1-14), with $F_{FVi} = 0$, $F'_{FVi} = 0$, and Eqs. (3.1-18), (3.1-28), and (3.1-20), becomes the same as that in the single phase convective heat transfer when Φ and Sc are replaced by Θ and Pr_V, respectively. In the case of a low mass transfer rate, we can define the local mass transfer coefficient β_x at the vapor-liquid interface and its Sherwood number Sh_x as

$$-\rho_V D\left(\frac{\partial W_{1V}}{\partial y}\right)_i = \beta_x(W_{1Vi} - W_{1V\infty}), \quad (3.5\text{-}1)$$

$$Sh_x = \frac{\beta_x x}{\rho_V D} = \left(\frac{\partial W_{1V}}{\partial y}\right)_i \frac{x}{(W_{1Vi} - W_{1V\infty})}. \quad (3.5\text{-}2)$$

In the above-mentioned situation we can derive the equations

$$Sh_x = (-\Phi'_{Fi})Re_V^{\frac{1}{2}} = C_F(Sc)Re_V^{\frac{1}{2}} \quad (3.5\text{-}3)$$

or

$$-\Phi'_{Fi} = C_F(Sc), \quad (3.5\text{-}4)$$

where $C_F(Sc)$ is a function of Sc with the same form as Eq. (3.4-12). Consequently, we can assume that $-\Phi'_{Fi}$ in the case of normal film thickness will be a function of $C_F(Sc)$.

Figure 3.5-1 shows the relation of $-\Phi'_{Fi}/C_F(Sc)$ versus W_R for the data in Table 3.3-1. The solid line in the figure represents the equation obtained by Fujii et al.:[1]

$$\frac{-\Phi'_{Fi}}{C_F(Sc)} = \left(\frac{2}{1 + W_R}\right)^{0.48} W_R. \quad (3.5\text{-}5)$$

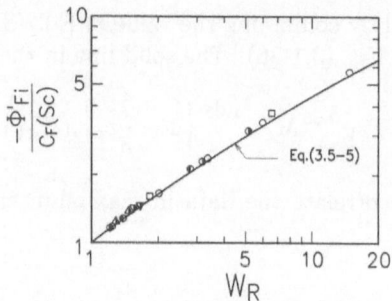

FIGURE 3.5-1. Correlation of $-\Phi'_{Fi}/C_F(Sc)$ versus W_R for the data in Table 3.3-1. Symbols used are the same as in Fig. 3.4-1

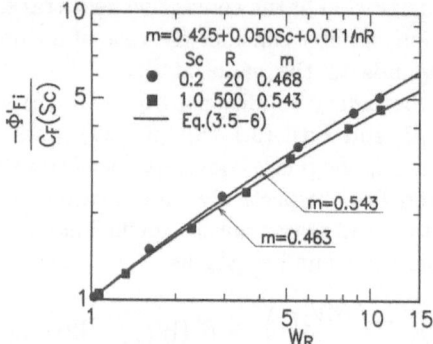

FIGURE 3.5-2. Correlation of $-\Phi'_{Fi}/C_F(Sc)$ versus W_R for the data in Table 3.3-2.

This equation is accurate for the case of $Sc = 1$ and $R \gtrsim 100$. However, the data for small values of both Sc and R are larger than the solid line by about ten percent.

Fujii et al.[4] correlated the data in Table 3.3-2 within two percent in maximum error by the equation

$$\frac{-\Phi'_{Fi}}{C_F(Sc)} = \left(\frac{2.5}{1.5 + W_R}\right)^m W_R, \qquad (3.5\text{-}6)$$

where

$$m = 0.425 + 0.050Sc + 0.011 \ln R \qquad (3.5\text{-}7)$$

in the ranges of $0.3 \lesssim Sc \lesssim 1$, $20 \lesssim R \lesssim 500$, and $1 \leq W_R \lesssim 10$. A few examples are shown in Fig. 3.5-2. Equation (3.5-6) can also correlate all the data in Table 3.3-1 within two percent in maximum error.

3.6 Algebraic Method for Calculating Condensation Mass Flux and Heat Flux

Combining Eqs. (3.4-10), (3.4-16), and (3.5-6) with the compatibility conditions (3.1-27) and (3.1-29) and the phase equilibrium relations (2-29) and (2-30), we can calculate algebraically and simultaneously the condensation mass flux, the heat flux, and the state at the vapor-liquid interface. Remember that the above three equations for Θ'_{FLi}, Θ'_{FVi}, and Φ'_{Fi} are valid independently of the compatibility conditions by referring to both the discussion on the number of boundary and compatibility conditions stated in Section 3.1 and the numerical solution shown in Table 3.3-2.

The substitution of Eqs. (3.4-10) and (3.4-16) into Eq. (3.1-27) yields

$$\underbrace{0.433\left(1.367 - \frac{0.432}{\sqrt{2\dot{M}_{FV}}} + \frac{1}{2\dot{M}_{FV}}\right)^{\frac{1}{2}}}_{①} \underbrace{(1+0.320H_i^{0.87})^{-1}}_{②}$$

$$= \underbrace{\frac{Pr_L \dot{M}_{FV}}{RH_i}}_{③} + \underbrace{C_F(Pr_V)\left(\frac{\lambda_V}{\lambda_L}\right)\left(\frac{\nu_L}{\nu_V}\right)^{\frac{1}{2}}\frac{(T_{V\infty}-T_i)}{(T_i-T_w)}}_{④}$$

$$\times \underbrace{\left[1+2.6Pr_V^{0.66}\dot{M}_{FV}^{1.05}\left\{1-\frac{2}{3}c_{P12}(W_{1Vi}-W_{1L})\right\}\right]}_{⑤}, \quad (3.6\text{-}1)$$

where the left-hand side term ① is the dimensionless heat flux in the condensate film at the vapor-liquid interface and term ② is appreciable only for a thick condensate film. Term ③ is the dimensionless latent heat released at the vapor-liquid interface, term ④ is the dimensionless heat flux of the vapor phase side, which is negligibly small for saturated vapors, and term ⑤ the effect of diffusion term in the vapor boundary layer, which is considerable for an air-water mixture.

The substitution of Eqs. (3.5-6) and (3.4-9) into Eq. (3.1-36), which is a substitute of Eq. (3.1-29), yields

$$\dot{M}_{FV} = \frac{C_F(Sc)}{Sc}\left(\frac{2.5}{1.5+W_R}\right)^m (W_R - 1). \quad (3.6\text{-}2)$$

The applicable ranges of Eqs. (3.6-1) and (3.6-2) are estimated at present as follows: $Pr_L=1$–15, $R=20$–500, $Pr_V =0.2$–1, $Sc=0.2$–1, and $\Delta T_{V\infty} \leq 100$ K.

The values of $\dot{M}_{FV}$, T_i, W_{1Vi}, and W_{1L} for given values of pressure p, temperature $T_{V\infty}$, and concentration $W_{1V\infty}$ in the bulk and cooling surface temperature T_w can be obtained by the following procedure: (1) Guess T_i;

(2) compute W_{1L} and W_{1Vi} using Eqs. (2-29) and (2-30), and compute representative physical properties and dimensionless numbers R, Pr_V, Sc, c_{p12}, and W_R; (3) obtain $\dot{M}_{FV}$ by substituting these values into Eq. (3.6-2); and (4) until the above values satisfy Eq. (3.6-1), modify the value T_i. The results thus obtained are named the algebraic solution.

All relevant physical quantities can be obtained from respective definition equation, e.g.,

$$\dot{m}_x = \frac{\mu_V}{x} \dot{M}_{FV} Re_{Vx}^{\frac{1}{2}}, \tag{3.6-3}$$

$$\dot{m}_{1x} = \dot{m}_x W_{1L}, \tag{3.6-4}$$

$$q_{wx} = 0.433 \left(1.367 - \frac{0.432}{\sqrt{2\dot{M}_{FV}}} + \frac{1}{2\dot{M}_{FV}}\right)^{\frac{1}{2}} \frac{\lambda_L}{x}(T_i - T_w) Re_{Lx}^{\frac{1}{2}}. \tag{3.6-5}$$

The values of $\dot{m}_x$ and q_{wx} obtained by the algebraic solution agree with those of the similarity solution within the errors of one percent and three percent for saturated vapors and superheated vapors of $\Delta T_{V\infty}=100$ K, respectively, as seen in Table 3.3-1.

The condensation mass flux $\bar{\dot{m}}$ and wall heat flux $\bar{q_w}$ averaged over the length $x = 0$ to ℓ are expressed as

$$\bar{\dot{m}} = 2(\dot{m}_x)_{x=\ell}, \tag{3.6-6}$$

$$\bar{q_w} = 2(q_{wx})_{x=\ell}. \tag{3.6-7}$$

3.7   Flow Resistance

The shear stresses τ_w and τ_i at the cooling surface and vapor-liquid interface are expressed, respectively, as

$$\tau_w = \mu_L \left(\frac{\partial U_L}{\partial y}\right)_w = \mu_L U_{V\infty} \left(\frac{U_{V\infty}}{\nu_L x}\right)^{\frac{1}{2}} F_{FLw}'' \tag{3.7-1a}$$

$$= \mu_V U_{V\infty} \left(\frac{U_{V\infty}}{\nu_V x}\right)^{\frac{1}{2}} R F_{FLw}'', \tag{3.7-1b}$$

$$\tau_i = \mu_V \left(\frac{\partial U_V}{\partial y}\right)_i = \mu_V U_{V\infty} \left(\frac{U_{V\infty}}{\nu_V x}\right)^{\frac{1}{2}} R F_{FLi}''. \tag{3.7-2}$$

Differentiating Eq. (3.2-1) twice with respect to η_{FL}, we obtain the following equation at $\eta_{FL} = \eta_{FLi}$:

$$F_{FLi}'' \approx F_{FLw}'' - \frac{F_{FLw}''^{\,2}}{12} \eta_{FLi}^3. \tag{3.7-3}$$

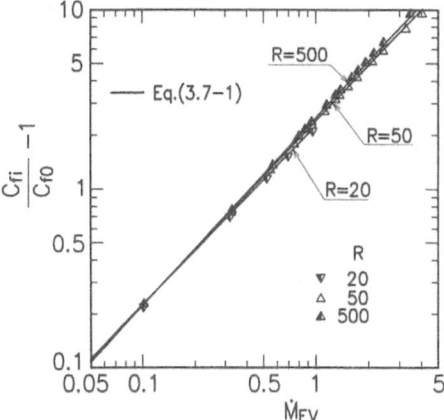

FIGURE 3.7-1. Correlation of $(C_{fi}/C_{f0})-1$ versus $\dot{M}_{FV}(=F_{FVi}/2)$.

Since the second term on the right-hand side of Eq. (3.7-3) is less than three percent of the first term in the range of numerical results in Tables 3.3-1 and 3.3-2, we can estimate that $F''_{FLi} \approx F''_{FLw}$, i.e., $\tau_i \approx \tau_w$. We now discuss the characteristics of τ_i.

The friction coefficient C_{fi} is defined by

$$C_{fi} = \frac{\tau_i}{\frac{1}{2}\rho_V U^2_{V\infty}} = 2F''_{FVi}\mathrm{Re}_{Vx}^{-\frac{1}{2}} . \tag{3.7-4}$$

When the condensation mass flux becomes extremely small, C_{fi} in Eq. (3.7-4) should asymptotically approach the following value:

$$C_{f0} = 0.664\mathrm{Re}_{Vx}^{-\frac{1}{2}} , \tag{3.7-5}$$

which is the friction coefficient for single phase forced convection, i.e.,

$$\frac{C_{fi}}{C_{f0}} = \frac{F''_{FVi}}{0.332} \to 1 . \tag{3.7-6}$$

In Fig. 3.7-1 the values of $\{(C_{fi}/C_{f0}) - 1\}$ are plotted against the corresponding dimensionless condensation mass flux $\dot{M}_{FV}$ with a parameter R from the numerical solution shown in Table 3.3-2. These data can be approximated by the equation[3]

$$\frac{C_{fi}}{C_{f0}} = 1 + \left(2.5 - \frac{5}{R}\right)\dot{M}_{FV}^{(1.05-\frac{1}{R})} \tag{3.7-7}$$

within the error of two percent in the range $0.1 \lesssim \dot{M}_{FV} \lesssim 5$ for $R = 50$ and 500, and $0.1 \lesssim \dot{M}_{FV} \lesssim 1$ for $R = 20$ as shown by solid lines in the figure.

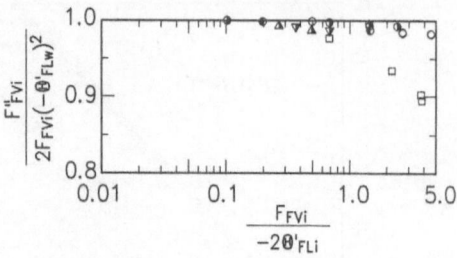

FIGURE 3.7-2.   Correlation of $F_{FVi}''/2F_{FVi}(-\Theta_{FLw}')^2$ versus $F_{FVi}/(-2\Theta_{FLi}')$. Symbols used are the same as in Fig. 3.4-1

Observation of the data in Table 3.3-1 reveals that the values of (F_{FVi}''/F_{FVi}) are nearly equal to those of $2(-\Theta_{FLw}')^2$. In Fig. 3.7-2 the ratios of these two values are plotted against $F_{FVi}/2(-\Theta_{FLi}')$. It is clear that the majority of the data lie within the ranges of 1 to 0.98 in the ordinate, particularly for the data for $F_{FVi}/2(-\Theta_{FLi}') < 0.2$, the following equation is valid:

$$\frac{F_{FVi}''}{2F_{FVi}(-\Theta_{FLw}')^2} = 1 . \tag{3.7-8}$$

For the case of $(T_{V\infty} - T_w) \approx 80$ K for a CFC114-CFC11 mixture, the value of the left-hand side of Eq. (3.7-8) decreases to 0.9. The magnitude and tendency of the data for RF_{FLw}'' is almost the same as in Fig. 3.7-2.

3.8   Relations between Relevant Physical Quantities and Dimensionless Functions

For condensate film:

$$U_L = U_{V\infty}F_{FL}' , \tag{3.8-1}$$

$$V_L = \frac{1}{2}\left(\frac{\nu_L U_{V\infty}}{x}\right)^{\frac{1}{2}} (\eta_{FL}F_{FL}' - F_{FL}) , \tag{3.8-2}$$

$$\frac{\partial U_L}{\partial x} = -\frac{U_{V\infty}}{2x}\eta_{FL}F_{FL}'' , \tag{3.8-3}$$

$$\frac{\partial U_L}{\partial y} = U_{V\infty}\left(\frac{U_{V\infty}}{\nu_L x}\right)^{\frac{1}{2}} F_{FL}'' , \tag{3.8-4}$$

$$\frac{\partial^2 U_L}{\partial y^2} = \frac{U_{V\infty}^2}{\nu_L x}F_{FL}''' , \tag{3.8-5}$$

$$\frac{\partial T_L}{\partial x} = -(T_i - T_w)\left(\frac{\eta_{FL}}{2x}\right)(-\Theta_{FL}') , \tag{3.8-6}$$

$$\frac{\partial T_L}{\partial y} = (T_i - T_w) \left(\frac{U_{V\infty}}{\nu_L x}\right)^{\frac{1}{2}} (-\Theta'_{FL}), \tag{3.8-7}$$

$$\frac{\partial^2 T_L}{\partial y^2} = (T_i - T_w) \left(\frac{U_{V\infty}}{\nu_L x}\right) (-\Theta''_{FL}), \tag{3.8-8}$$

$$\delta = \left(\frac{\nu_L x}{U_{V\infty}}\right)^{\frac{1}{2}} \eta_{FLi}, \tag{3.8-9}$$

$$\frac{d\delta}{dx} = \frac{1}{2} \left(\frac{\nu_L}{U_{V\infty}x}\right)^{\frac{1}{2}} \eta_{FLi} = \frac{\delta}{2x}, \tag{3.8-10}$$

$$\dot{m}_x = \rho_L \left(U_L \frac{d\delta}{dx} - V_L\right)_i = \left(\frac{\mu_L}{2x}\right) \left(\frac{U_{V\infty}x}{\nu_L}\right)^{\frac{1}{2}} F_{FLi}. \tag{3.8-11}$$

For the vapor boundary layer:

$$U_V = U_{V\infty} F'_{FV}, \tag{3.8-12}$$

$$V_V = \frac{1}{2} \left(\frac{\nu_V U_{V\infty}}{x}\right)^{\frac{1}{2}} (\eta_{FV} F'_{FV} - F_{FV}) + U_{V\infty} F'_{FV} \frac{d\delta}{dx}, \tag{3.8-13}$$

$$\frac{\partial U_V}{\partial x} = -U_{V\infty} F''_{FV} \left\{ \left(\frac{U_{V\infty}}{\nu_V x}\right) \frac{d\delta}{dx} + \frac{\eta_{FV}}{2x} \right\}, \tag{3.8-14}$$

$$\frac{\partial U_V}{\partial y} = U_{V\infty} \left(\frac{U_{V\infty}}{\nu_V x}\right)^{\frac{1}{2}} F''_{FV}, \tag{3.8-15}$$

$$\frac{\partial^2 U_V}{\partial y^2} = \frac{U_{V\infty}^2}{\nu_V x} F'''_{FV}, \tag{3.8-16}$$

$$\frac{\partial T_V}{\partial x} = (T_{V\infty} - T_i) \left\{ \left(\frac{U_{V\infty}}{\nu_V x}\right)^{\frac{1}{2}} \frac{d\delta}{dx} + \frac{\eta_{FV}}{2x} \right\} (-\Theta'_{FV}), \tag{3.8-17}$$

$$\frac{\partial T_V}{\partial y} = (T_{V\infty} - T_i) \left(\frac{U_{V\infty}}{\nu_V x}\right)^{\frac{1}{2}} (-\Theta'_{FV}), \tag{3.8-18}$$

$$\frac{\partial^2 T_V}{\partial y^2} = (T_{V\infty} - T_i) \left(\frac{U_{V\infty}}{\nu_V x}\right) (-\Theta''_{FV}), \tag{3.8-19}$$

$$\frac{\partial W_{1V}}{\partial y} = (W_{1Vi} - W_{1V\infty}) \left(\frac{U_{V\infty}}{\nu_V x}\right)^{\frac{1}{2}} \Phi'_F, \tag{3.8-20}$$

$$\frac{\partial^2 W_{1V}}{\partial y^2} = (W_{1Vi} - W_{1V\infty}) \left(\frac{U_{V\infty}}{\nu_V x}\right) \Phi''_F, \tag{3.8-21}$$

$$\dot{m}_x = \rho_V \left(U_V \frac{d\delta}{dx} - V_V\right)_i = \left(\frac{\mu_V}{2x}\right) \left(\frac{U_{V\infty}x}{\nu_V}\right)^{\frac{1}{2}} F_{FVi}. \tag{3.8-22}$$

REFERENCES

1. Fujii, T., H. Uehara, K. Mihara, and T. Kato, Forced Convection Condensation in the Presence of Noncondensables—A Theoretical Treatment for Two-Phase Laminar Boundary Layer (in Japanese), *Reports of Research Institute of Industrial Science, Kyushu University*, No. 66, 53–80 (1977).

2. Rose, J. W., Boundary-Layer Flow on a Flat Plate, *Int. J. Heat Mass Transfer*, 22, 969 (1979).

3. Fujii, T., Sh. Koyama and M. Watabe, Laminar Forced-Convection Condensation of Binary Mixtures on a Flat Plate (in Japanese), *Trans. Jpn. Soc. Mech. Eng.*, 56, 486, 541–548 (1987).

4. Fujii, T., J. B. Lee, K. Shinzato, and M. Watabe, A Correlation Equation for Mass Concentration Gradient at the Vapor-Liquid Interface in Forced-Convection Condensation of a Binary Vapor on a Vertical Surface (in preparation).

4

Similarity Solution for Free-Convection Condensation of Binary Vapors

The basic equations, boundary conditions, and compatibility conditions for free-convection condensation of a binary vapor mixture correspond to Eqs. (2.1)–(2.7), (2.14)–(2-19), and (2-20)–(2-27), respectively, where $U_{V\infty}$ in Eq. (2-17) is set to zero. Prior to performing similarity transformation to these equations, it is necessary to express $\rho_{V\infty}/\rho_V$ in Eq. (2-5) as a function of concentration and temperature. Since the total pressure is invariable through the vapor boundary layer, the following equation is valid under the assumption that each component vapor is an ideal gas:

$$\frac{\rho_{V\infty}}{\rho_V} = \frac{M_{V\infty}T_V}{M_V T_{V\infty}} , \qquad (4\text{-}1)$$

where

$$M_V = \frac{M_1 M_2}{M_1 - (M_1 - M_2)W_{1V}} . \qquad (4\text{-}2)$$

Using Eqs. (4-1) and (4.2), we can derive the buoyancy term (divided by g) as

$$1 - \frac{\rho_{V\infty}}{\rho_V} = \Omega_W(W_{1V} - W_{1V\infty}) + \Omega_T(T_{V\infty} - T_V)$$
$$- \Omega_W\Omega_T(W_{1V} - W_{1V\infty})(T_{V\infty} - T_V) , \qquad (4\text{-}3)$$

where

$$\Omega_W = \frac{M_1 - M_2}{M_1 - (M_1 - M_2)W_{1V\infty}} , \qquad (4\text{-}4)$$

$$\Omega_T = \frac{1}{T_{V\infty}} . \qquad (4\text{-}5)$$

The relation $W_{1V} > W_{1V\infty}$ is valid in the vapor boundary layer under the rule of the subscripts 1 and 2 defined in Chapter 2, and $\Omega_W > 0$ for the case of $M_1 > M_2$. In this case the direction of buoyancy is positive (x direction) because the second and third terms on the right-hand side of Eq. (4-3) are usually smaller than the first term, and then assumption (2) in Chapter 2 is valid. In the case of $M_1 < M_2$, so far as the value of $\Omega_T(T_{V\infty} - T_V)$ is not very large, the direction of buoyancy is negative, and the vapor flows in the reverse direction of the condensate, and consequently similarity solution does not exist.

Since the mathematical treatment of free-convection condensation in this chapter is similar in principle to that of forced-convection condensation in Chapter 3, duplicate description will be avoided. The physical quantities for free-convection condensation are denoted with the subscript G.

4.1   Similarity Transformation

We define the dimensionless variables η_{GL} and η_{GV} and dimensionless stream functions $F_{GL}(\eta_{GL})$ and $F_{GV}(\eta_{GV})$ by the following equations, respectively:

$$\eta_{GL} = y \left(\frac{g}{4\nu_L^2 x} \right)^{\frac{1}{4}}, \tag{4.1-1}$$

$$\eta_{GV} = (y - \delta) \left(\frac{g}{4\nu_V^2 x} \right)^{\frac{1}{4}}, \tag{4.1-2}$$

$$F_{GL}(\eta_{GL}) = \frac{\Psi_L(x,y)}{2\sqrt{2}(g\nu_L^2 x^3)^{\frac{1}{4}}}, \tag{4.1-3}$$

$$F_{GV}(\eta_{GV}) = \frac{\Psi_V(x,y)}{2\sqrt{2}(g\nu_V^2 x^3)^{\frac{1}{4}}}, \tag{4.1-4}$$

and define the dimensionless temperatures $\Theta_{GL}(\eta_{GL})$ and $\Theta_{GV}(\eta_{GV})$ and normalized concentration $\Phi_G(\eta_{GV})$ by the functions of the same form as Eqs. (3.1-7), (3.1-8), and (3.1-9), respectively. The transformation of Eqs. (2-1)–(2-7) by using these equations yields

$$F_{GL}''' + 3F_{GL}''F_{GL} - 2F_{GL}'^2 + 1 = 0, \tag{4.1-5}$$

$$\Theta_{GL}'' + 3Pr_L F_{GL}\Theta_{GL}' = 0, \tag{4.1-6}$$

$$F_{GV}''' + 3F_{GV}''F_{GV} - 2F_{GV}'^2 + \omega_W\Phi_G + \omega_T\Theta_{GV} - \omega_W\omega_T\Phi_G\Theta_{GV} = 0, \tag{4.1-7}$$

$$\Theta_{GV}'' + 3Pr_V F_{GV}\Theta_{GV}' + \frac{Pr_V}{Sc}c_{p12}(W_{1Vi} - W_{1V\infty})\Phi_G'\Theta_{GV}' = 0, \tag{4.1-8}$$

$$\Phi_G'' + 3ScF_{GV}\Phi_G' = 0, \tag{4.1-9}$$

where the prime denotes the differential derivative with respect to η_{GL} or η_{GV} and

$$\omega_W = \Omega_W(W_{1Vi} - W_{1V\infty}), \tag{4.1-10}$$

$$\omega_T = \Omega_T(T_{V\infty} - T_i). \tag{4.1-11}$$

The boundary conditions (2-14)–(2-19) are transformed as follows at $\eta_{GL} = 0$:

$$F'_{GLw} = 0 , \tag{4.1-12}$$
$$F_{GLw} = 0 , \tag{4.1-13}$$
$$\Theta_{GLw} = 1 ; \tag{4.1-14}$$

as $\eta_{GV} \to \infty$:

$$F'_{GV\infty} = 0 , \tag{4.1-15}$$
$$\Theta_{GV\infty} = 0 , \tag{4.1-16}$$
$$\Phi_{G\infty} = 0 ; \tag{4.1-17}$$

and the compatibility conditions (2-20)–(2-27) are transformed as follows at $\eta_{GL} = \eta_{GLi}$ or $\eta_{GV} = 0$:

$$F'_{GVi} = F'_{GLi} , \tag{4.1-18}$$
$$F''_{GVi} = RF''_{GLi} , \tag{4.1-19}$$
$$F_{GVi} = RF_{GLi} \tag{4.1-20}$$
$$= \frac{1}{3}R(\dot{M}_{1GL} + \dot{M}_{GL}) = \frac{1}{3}R\dot{M}_{GL} , \tag{4.1-21}$$
$$\Theta_{GLi} = 0 , \tag{4.1-22}$$
$$\Theta_{GVi} = 1 , \tag{4.1-23}$$
$$-\Theta'_{GLi} = \frac{\mathrm{Pr}_L \dot{M}_{GL}}{H_i} + \frac{\lambda_V}{\lambda_L}\left(\frac{\nu_L}{\nu_V}\right)^{\frac{1}{2}} \frac{(T_{V\infty} - T_i)}{(T_i - T_w)}(-\Theta'_{GVi}) , \tag{4.1-24}$$
$$\Phi_{Gi} = 1 , \tag{4.1-25}$$
$$-\Phi'_{Gi} = \frac{W_{1Vi}\dot{M}_{2GL} - (1 - W_{1Vi})\dot{M}_{1GL}}{W_{1Vi} - W_{1V\infty}}R\mathrm{Sc} , \tag{4.1-26}$$

where the dimensionless condensation mass flux $\dot{M}_{kGL}$ for the component $k(= 1, 2)$ and the dimensionless total condensation mass flux $\dot{M}_{GL}$ are given by

$$\dot{M}_{kGL} = \frac{\dot{m}_{kx}x}{\mu_L}\left(\frac{Ga_x}{4}\right)^{-\frac{1}{4}} , \tag{4.1-27}$$
$$\dot{M}_{GL} = \frac{\dot{m}_x x}{\mu_L}\left(\frac{Ga_x}{4}\right)^{-\frac{1}{4}} , \tag{4.1-28}$$

where Ga_x is the Galileo number defined by

$$Ga_x = \frac{gx^3}{\nu_L^2} . \tag{4.1-29}$$

Since $\dot{M}_{kGL}$ should be independent of x for the sake of an existence of similarity solution, $\dot{m}_x$ is proportional to $x^{-1/4}$.

Equation (2-28) is transformed as

$$W_{kL} = \frac{\dot{M}_{kGL}}{\dot{M}_{GL}} \qquad (k = 1, 2) \tag{4.1-30}$$

and the equation corresponding to Eq. (3.1-36) in the case of forced-convection condensation is expressed as

$$1 - \frac{3Sc F_{GVi}}{(-\Phi'_{Gi})} = \frac{1}{W_R}. \tag{4.1-31}$$

This is an alternative to Eq. (4.1-26).

4.2   Procedure of Numerical Calculation

Though Eqs. (4.1-7), (4.1-8), and (4.1-9) depend upon each other, the procedure of numerical calculation is almost the same as in the case of forced-convection condensation. The procedure is briefly explained by using the flow chart of Fig. 4.2-1.

Procedures (1) to (4) are the same as those in the case of forced-convection condensation.

(5) Guess a boundary value of F''_{GLW} and a calculation domain $\eta_{GV\infty}$.

(6) Compute the value of F_{GL} from the following equation, which is obtained by solving Eq. (4.1-5) subject to Eqs. (4.1-12) and (4.1-13), by means of the series expansion method :

$$\begin{aligned}
F_{GL} &= \frac{F''_{GLw}}{2}\eta_{GL}^2 - \frac{1}{3!}\eta_{GL}^3 + \frac{4}{5!}\left(\frac{F''_{GLw}}{2}\right)^2 \eta_{GL}^5 \\
&\quad - \frac{104}{8!}\left(\frac{F''_{GLw}}{2}\right)^3 \eta_{GL}^8 - \frac{72}{9!}\left(\frac{F''_{GLw}}{2}\right)^3 \eta_{GL}^9 \\
&\quad + \frac{13680}{11!}\left(\frac{F''_{GLw}}{2}\right)^4 \eta_{GL}^{11} + \cdots .
\end{aligned} \tag{4.2-1}$$

(7) Numerically solve Eq. (4.1-7) subject to the boundary values F_{GVi}, F'_{GVi}, and F''_{GVi}, which are obtained from Eq. (4.2-1) together with Eqs. (4.1-18)–(4.1-20), by means of the Runge-Kutta-Verner method. In this case we take $\Phi_G = 0$ and $\Theta_{GV} = 0$ at first, and give the prior values of Φ_G and Θ_{GV} in repetition.

(8) Compute Φ_G by substituting the previously obtained F_{GV} into the

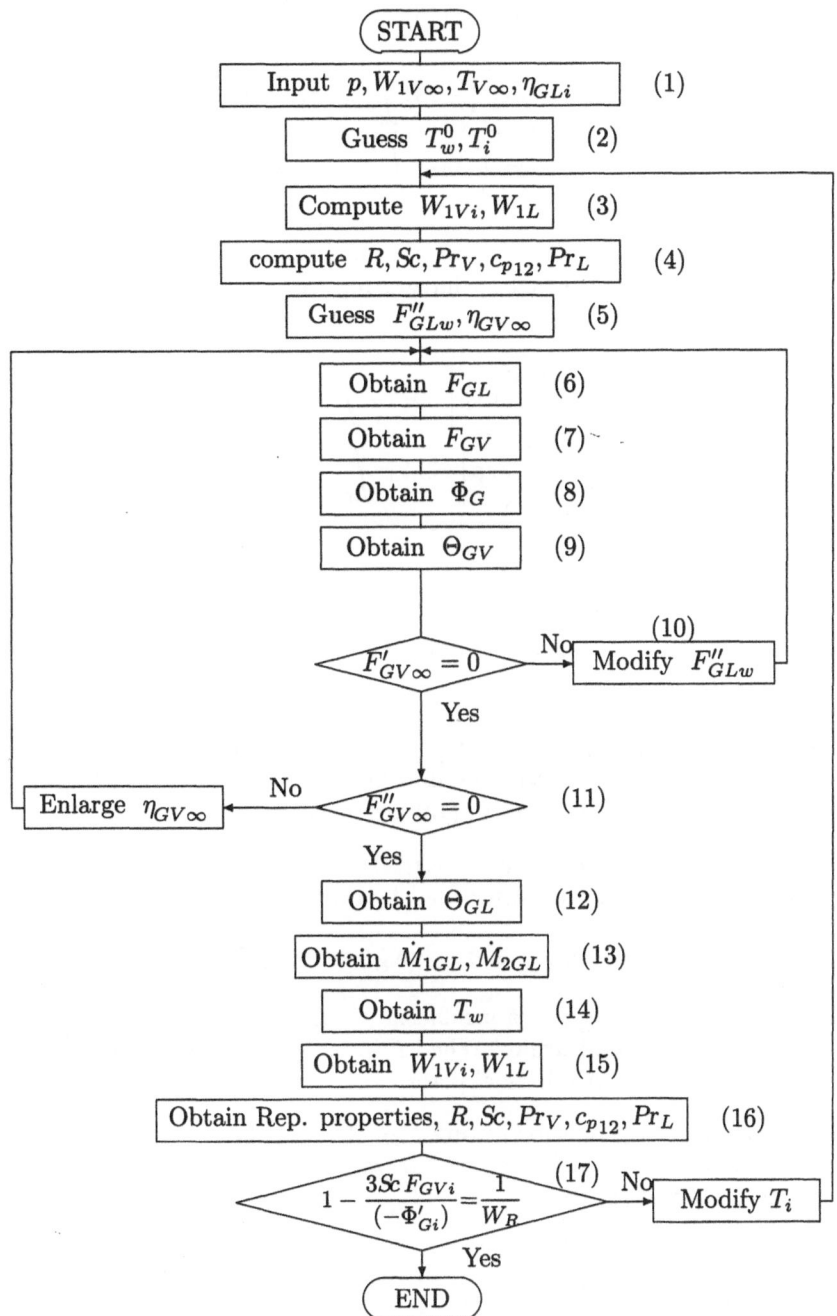

FIGURE 4.2-1. Flow chart for numerically solving the equation system of free-convection condensation of a binary vapor mixture.

following equation, which is obtained by integrating Eq. (4.1-9) subject to Eqs. (4.1-17) and (4.1-25),

$$
\Phi_G = 1 - \frac{\int_0^{\eta_{GV}} \exp\left(-3Sc \int_0^{\eta_{GV}} F_{GV} d\eta_{GV}\right) d\eta_{GV}}{\int_0^{\infty} \exp\left(-3Sc \int_0^{\eta_{GV}} F_{GV} d\eta_{GV}\right) d\eta_{GV}} . \tag{4.2-2}
$$

(9) Compute Θ_{GV} by substituting F_{GV} and Φ_G into the following equation, which is obtained by integrating Eq. (4.1-8) subject to Eqs. (4.1-16) and (4.1-23),

$$
\Theta_{GV} = 1 - \frac{\int_0^{\eta_{GV}} \exp\left[-3Pr_V \int_0^{\eta_{GV}} \left\{ F_{GV} + \frac{c_{p12}}{3Sc}(W_{1Vi} - W_{1V\infty})\Phi_G \right\} d\eta_{GV} \right] d\eta_{GV}}{\int_0^{\infty} \exp\left[-3Pr_V \int_0^{\eta_{GV}} \left\{ F_{GV} + \frac{c_{p12}}{3Sc}(W_{1Vi} - W_{1V\infty})\Phi_G \right\} d\eta_{GV} \right] d\eta_{GV}} . \tag{4.2-3}
$$

(10) Modify the assumed value F''_{GLw} until the condition (4.1-15) is satisfied within a convergence radius by means of the Newton-Raphson method, where Steps (6) to (10) are repeated.

(11) After the value F''_{GLw} is converged, check whether the calculation domain $\eta_{GV\infty}$ is sufficient or not by the equation

$$
F''_{GV\infty} = 0 \qquad (\text{e.g., } < 10^{-5}). \tag{4.2-4}
$$

If the value of $\eta_{GV\infty}$ is insufficient, it is enlarged and Steps (6) to (11) are repeated again. When $F''_{GV\infty} < 10^{-5}$, the value of F''_{GLw} is fixed to the fourth decimal place.

(12) Compute Θ_{GL} by substituting F_{GL} of Eq. (4.2-1) into the following equation, which is derived by integrating Eq. (4.1-6) subject to Eqs. (4.1-14) and (4.1-22),

$$
\Theta_{GL} = 1 - \frac{\int_0^{\eta_{GL}} \exp\left(-3Pr_L \int_0^{\eta_{GL}} F_{GL} d\eta_{GL}\right) d\eta_{GL}}{\int_0^{\eta_{GLi}} \exp\left(-3Pr_L \int_0^{\eta_{GL}} F_{GL} d\eta_{GL}\right) d\eta_{GL}} . \tag{4.2-5}
$$

(13) Compute $\dot{M}_{1GL}$ and $\dot{M}_{2GL}$ by using Eqs. (4.1-21) and (4.1-30).

(14) Compute T_w from Eq. (4.1-24).

(15) Compute W_{1Vi} and W_{1L} from Eqs. (2-30) and (2-29).

(16) Compute the relevant physical properties and dimensionless number R, Pr_L, Pr_V, Sc, and c_{p12}.

(17) Substitute the obtained values of F_{GVi}, $-\Phi_{Gi}$, and W_R into Eq. (4.1-31). If it is not satisfied, replace the guessed value $T_i{}^\circ$ with a new value T_i. Then, return to Step (3), and repeat the successive computation. Thus, T_i and T_w are determined together with the other boundary values.

4.3   Examples of Numerical Solutions

4.3.1   CHARACTERISTICS OF THE BOUNDARY VALUES

There are a few mixtures which satisfy the conditions of existence of similarity solution. Table 4.3-1 shows the boundary values, relevant dimensionless numbers, heat fluxes, and condensation mass flux, which have been numerically solved under specified conditions of p, $T_{V\infty}$, $W_{1V\infty}$, and η_{GLi} for mixtures of ethanol-water, CFC114-CFC11, and air-water.

From Table 4.3-1 we can find some characteristics as follows:

1. The differences between the values of F''_{GLw}, $-\Theta'_{GLw}$, and $-\Theta'_{GLi}$ and those of the Nusselt theory, i.e., $F''_{GLw} = \eta_{GLi}$ and $-\Theta'_{GLw} = -\Theta'_{GLi} = \eta_{GLi}{}^{-1}$(see Section 5.3), are in maximum 4 percent, 0.6 percent, and 2 percent, respectively.

2. In the case of saturated vapors, the buoyancy force due to temperature difference ω_T increases as η_{GLi} increases, and the ratios ω_T/ω_W for the mixtures of ethanol-water and air-water reach about 0.1–0.2 in maximum, while the ratio ω_T/ω_W for the CFC114-CFC11 mixture ($M_1/M_2 = 1.24$) reaches 0.3 in maximum. Accordingly, the term ω_T must not be neglected for a large value of η_{GLi} and for the mixture whose value of M_1/M_2 is near unity, though only the buoyancy force due to concentration difference ω_W has been discussed in most reported papers.

3. In the case of the vapors with large degree of superheat, the value of ω_T is of the same order of, or larger than, the value of ω_W.

4. The convective heat flux q_{cx} at the vapor-liquid interface in the vapor phase is less than three percent of the heat flux q_{wx} at the cooling

TABLE 4.3-1. Boundary values, dimensionless numbers, heat flux, and condensation mass flux for free-convection condensation of binary vapor mixtures.

No.	$\frac{p}{\text{MPa}}$	$\frac{T_{s\infty}}{°C}$	$W_{1V\infty}$	η_{GLi}	$\frac{\Delta T_{V\infty}}{°C}$	$\frac{T_i}{°C}$	$\frac{T_w}{°C}$	W_{1Vi}	W_{1L}	R	Sc	Pr_V	Pr_L	$-c_{p12}$
1- 1	0.1	98.0	0.1451	0.15	0	94.26	93.76	0.3818	0.0545	176.1	0.7305	0.9222	1.8353	0.101
1- 2	0.1	98.0	0.1451	0.15	0	94.26	93.76	0.3818	0.0545	176.1	0.7305	0.9222	1.8353	0
1- 3	0.1	98.0	0.1451	0.15	100	94.86	94.31	0.3503	0.0472	175.9	0.7393	0.9252	1.7883	0.101
1- 4	0.1	98.0	0.1451	0.30	0	88.76	75.04	0.5957	0.1432	210.3	0.6701	0.9024	3.2506	0.101
1- 5	0.1	98.0	0.1451	0.30	0	88.76	75.05	0.5957	0.1432	210.3	0.6701	0.9024	3.2504	0
1- 6	0.1	98.0	0.1451	0.30	100	88.76	73.30	0.5957	0.1431	213.3	0.6702	0.9025	3.3397	0.101
1- 7	0.1	94.0	0.3946	0.15	0	85.12	84.14	0.6916	0.2487	194.9	0.5708	0.8752	3.8252	0.101
1- 8	0.1	94.0	0.3946	0.25	0	82.63	71.11	0.7472	0.3907	221.2	0.5546	0.8715	6.2017	0.101
1- 9	0.1	88.5	0.6034	0.20	0	80.77	75.33	0.7910	0.5878	204.9	0.4803	0.8585	8.0369	0.101
1-10	0.1	88.5	0.6034	0.20	100	80.80	74.54	0.7902	0.5845	206.3	0.4805	0.8585	8.0956	0.101
1-11	0.1	88.5	0.6034	0.25	0	80.66	64.85	0.7941	0.6014	223.7	0.4794	0.8584	9.4957	0
1-12	0.1	88.5	0.6034	0.25	0	80.66	64.85	0.7941	0.6014	223.7	0.4794	0.8584	9.4961	0.101
1-13	0.1	88.5	0.6034	0.25	100	80.66	61.87	0.7940	0.6012	229.7	0.4794	0.8584	9.9262	0.101
2- 1	0.5	75.0	0.2189	0.16	0	74.41	74.02	0.2527	0.1576	32.2	0.4870	0.7973	3.7762	-0.229
2- 2	0.5	75.0	0.2189	0.16	0	74.41	74.02	0.2527	0.1576	32.2	0.4870	0.7973	3.7763	0
2- 3	0.5	75.0	0.2189	0.16	100	74.67	73.50	0.2382	0.1475	32.2	0.4871	0.7989	3.7618	-0.230
2- 4	0.5	75.0	0.2189	0.28	0	73.39	69.63	0.3085	0.1976	32.5	0.4867	0.7917	3.8643	-0.228
2- 5	0.5	75.0	0.2189	0.28	0	73.39	69.63	0.3085	0.1976	32.5	0.4867	0.7917	3.8643	0
2- 6	0.5	75.0	0.2189	0.28	100	73.67	67.42	0.2935	0.1866	32.8	0.4868	0.7932	3.8577	-0.228
2- 7	0.5	65.0	0.6819	0.14	0	64.12	63.88	0.7148	0.5758	30.7	0.4556	0.7501	4.5286	0
2- 8	0.5	65.0	0.6819	0.14	0	64.12	63.88	0.7148	0.5758	30.7	0.4556	0.7501	4.5286	-0.204
2- 9	0.5	65.0	0.6819	0.22	0	62.97	61.55	0.7545	0.6240	30.7	0.4529	0.7500	4.6224	-0.203
2-10	0.5	58.0	0.9059	0.15	0	57.41	57.11	0.9226	0.8642	29.3	0.4181	0.7587	5.0456	-0.195
2-11	0.5	58.0	0.9059	0.15	100	57.67	56.33	0.9157	0.8529	29.5	0.4188	0.7584	5.0325	-0.195
2-12	0.5	58.0	0.9059	0.22	0	56.94	55.57	0.9350	0.8848	29.4	0.4167	0.7593	5.0892	0
2-13	0.5	58.0	0.9059	0.22	0	56.94	55.57	0.9350	0.8848	29.4	0.4167	0.7593	5.0892	-0.194
2-14	0.5	58.0	0.9059	0.22	100	57.21	53.86	0.9280	0.8730	29.7	0.4175	0.7590	5.0825	-0.194
2-15	0.8	82.0	0.7722	0.20	0	80.67	79.83	0.8116	0.7071	21.1	0.4374	0.7874	4.6752	-0.208
2-16	0.8	82.0	0.7722	0.30	0	79.70	75.47	0.8399	0.7458	21.3	0.4344	0.7874	4.7468	-0.207
2-17	0.8	90.0	0.4767	0.25	0	88.01	85.89	0.5572	0.4174	22.4	0.4757	0.8018	4.1699	-0.231
2-18	0.8	90.0	0.4767	0.35	0	87.03	78.82	0.5959	0.4558	22.8	0.4741	0.7999	4.2463	-0.229
3- 1	0.1	98.5	0.0625	0.16	0	93.45	92.77	0.2894	0	188.2	0.5240	0.8700	1.8808	0.545
3- 2	0.1	98.5	0.0625	0.20	0	78.42	76.34	0.6682	0	190.6	0.5359	0.7943	2.3068	0
3- 3	0.1	98.5	0.0625	0.20	0	78.11	76.01	0.6732	0	190.8	0.5361	0.7935	2.3171	0.601
3- 4	0.1	96.0	0.1842	0.14	0	87.69	87.26	0.4734	0	181.3	0.5335	0.8066	2.0109	0.589
3- 5	0.1	96.0	0.1842	0.14	0	87.71	87.28	0.4728	0	181.3	0.5335	0.8067	2.0103	0
3- 6	0.1	96.0	0.1842	0.14	100	89.53	89.06	0.4211	0	181.5	0.5319	0.8160	1.9668	0.581
3- 7	0.1	96.0	0.1842	0.14	100	89.61	89.14	0.4187	0	181.6	0.5318	0.8165	1.9648	0
3- 8	0.1	96.0	0.1842	0.16	0	79.37	78.54	0.6523	0	183.4	0.5394	0.7781	2.2480	0.618
3- 9	0.1	96.0	0.1842	0.16	0	79.50	78.67	0.6501	0	183.3	0.5394	0.7784	2.2440	0
3-10	0.1	96.0	0.1842	0.16	100	82.72	81.83	0.5898	0	182.3	0.5373	0.7875	2.1512	0.608
3-11	0.1	96.0	0.1842	0.16	100	83.04	82.17	0.5831	0	182.2	0.5371	0.7885	2.1418	0
3-12	0.1	90.0	0.4068	0.12	0	75.50	75.21	0.7126	0	174.4	0.5495	0.7430	2.3579	0.669
3-13	0.1	90.0	0.4068	0.13	0	67.25	66.80	0.8081	0	180.2	0.5531	0.7336	2.6667	0.687
3-14	0.1	90.0	0.4068	0.13	0	67.49	67.05	0.8058	0	180.0	0.5530	0.7338	2.6567	0
3-15	0.1	90.0	0.4068	0.13	10	68.13	67.69	0.7996	0	179.5	0.5528	0.7344	2.6307	0.686
3-16	0.1	90.0	0.4068	0.13	10	68.42	67.98	0.7967	0	179.3	0.5527	0.7346	2.6192	0
3-17	0.1	90.0	0.4068	0.13	100	72.47	72.00	0.7521	0	176.3	0.5510	0.7390	2.4670	0.676
3-18	0.1	90.0	0.4068	0.13	100	72.85	72.39	0.7475	0	176.1	0.5508	0.7394	2.4531	0

TABLE 4.3-1.(2)

No.	$C_G(Sc)$	$C_G(Pr_V)$	ω_W	ω_T	ω_i	χ_i	χ_i^+	$\dfrac{M_{GL}}{10^{-2}}$	$\dot{M}_{GV}$	W_R
1- 1	0.3873	0.3971	0.15808	0.01008	0.16656	0.16545	0.18609	0.3405	0.9390	3.610
1- 2	0.3873	0.3971	0.15807	0.01008	0.16655	0.16544	0.18608	0.3405	0.9390	3.610
1- 3	0.3878	0.3972	0.13704	0.21890	0.32595	0.30272	0.34221	0.3427	0.7976	3.096
1- 4	0.3836	0.3962	0.30097	0.02489	0.31837	0.31493	0.36666	2.6847	7.5150	234.273
1- 5	0.3836	0.3962	0.30097	0.02489	0.31837	0.31493	0.36666	2.6847	7.5148	234.261
1- 6	0.3836	0.3962	0.30091	0.23185	0.46300	0.43094	0.51127	2.6867	6.9447	227.033
1- 7	0.3764	0.3949	0.23803	0.02418	0.25646	0.25180	0.31317	0.3417	0.9361	3.035
1- 8	0.3751	0.3948	0.28259	0.03098	0.30482	0.29855	0.37649	1.5627	4.6516	89.595
1- 9	0.3685	0.3941	0.18053	0.02137	0.19805	0.19266	0.25888	0.8043	2.4697	12.996
1-10	0.3685	0.3941	0.17982	0.23330	0.37117	0.31241	0.43171	0.8069	2.1333	10.866
1-11	0.3684	0.3941	0.18351	0.02169	0.20122	0.19574	0.26328	1.5609	5.2150	95.078
1-12	0.3684	0.3941	0.18351	0.02169	0.20122	0.19574	0.26328	1.5608	5.2151	95.106
1-13	0.3684	0.3941	0.18347	0.23360	0.37421	0.31518	0.43625	1.5632	4.5916	86.592
2- 1	0.3692	0.3910	0.00693	0.00169	0.00860	0.00823	0.01054	0.4114	0.4343	1.551
2- 2	0.3692	0.3910	0.00693	0.00169	0.00860	0.00824	0.01054	0.4114	0.4343	1.551
2- 3	0.3692	0.3911	0.00396	0.22388	0.22695	0.17788	0.22806	0.4345	0.2030	1.271
2- 4	0.3691	0.3907	0.01837	0.00463	0.02291	0.02191	0.02797	2.1935	1.8317	5.219
2- 5	0.3691	0.3907	0.01837	0.00463	0.02291	0.02191	0.02797	2.1935	1.8317	5.221
2- 6	0.3691	0.3908	0.01529	0.22610	0.23793	0.18896	0.24216	2.2527	1.0578	3.312
2- 7	0.3661	0.3884	0.00745	0.00262	0.01005	0.00947	0.01215	0.2762	0.2678	1.310
2- 8	0.3661	0.3884	0.00745	0.00262	0.01005	0.00947	0.01216	0.2762	0.2678	1.310
2- 9	0.3658	0.3884	0.01642	0.00600	0.02232	0.02098	0.02703	1.0706	0.8514	2.253
2-10	0.3620	0.3889	0.00399	0.00178	0.00576	0.00530	0.00715	0.3387	0.3605	1.401
2-11	0.3621	0.3889	0.00233	0.23271	0.23450	0.17471	0.23530	0.3624	0.1534	1.184
2-12	0.3619	0.3890	0.00695	0.00319	0.01012	0.00929	0.01255	1.0662	0.9888	2.379
2-13	0.3619	0.3890	0.00695	0.00319	0.01012	0.00929	0.01255	1.0662	0.9888	2.378
2-14	0.3620	0.3889	0.00526	0.23377	0.23781	0.17741	0.23964	1.1152	0.4736	1.671
2-15	0.3642	0.3905	0.00911	0.00375	0.01283	0.01187	0.01594	0.8048	0.5052	1.606
2-16	0.3638	0.3905	0.01565	0.00647	0.02201	0.02035	0.02743	2.6964	1.4923	3.564
2-17	0.3681	0.3913	0.01742	0.00548	0.02280	0.02154	0.02800	1.5706	0.9037	2.356
2-18	0.3679	0.3912	0.02578	0.00819	0.03376	0.03187	0.04146	4.2534	2.2617	6.690
3- 1	0.3725	0.3947	0.08788	0.01360	0.10028	0.09724	0.12564	0.4118	1.3771	4.631
3- 2	0.3736	0.3909	0.23460	0.05403	0.27596	0.26631	0.32697	0.8063	2.1208	10.693
3- 3	0.3736	0.3908	0.23653	0.05486	0.27841	0.26864	0.32965	0.8062	2.1180	10.773
3- 4	0.3734	0.3915	0.11752	0.02253	0.13740	0.13319	0.16438	0.2768	0.8246	2.569
3- 5	0.3734	0.3915	0.11729	0.02247	0.13713	0.13293	0.16407	0.2768	0.8250	2.566
3- 6	0.3732	0.3920	0.09628	0.22694	0.30137	0.25765	0.32435	0.2788	0.6829	2.286
3- 7	0.3732	0.3920	0.09529	0.22677	0.30045	0.25669	0.32323	0.2788	0.6836	2.272
3- 8	0.3739	0.3900	0.19025	0.04505	0.22672	0.21919	0.26497	0.4142	1.1007	3.540
3- 9	0.3739	0.3900	0.18935	0.04470	0.22559	0.21810	0.26371	0.4142	1.1018	3.528
3-10	0.3737	0.3905	0.16483	0.24147	0.36649	0.32449	0.40120	0.4161	0.9745	3.201
3-11	0.3737	0.3906	0.16212	0.24078	0.36386	0.32181	0.39817	0.4162	0.9760	3.165
3-12	0.3747	0.3880	0.13663	0.03994	0.17112	0.16553	0.19337	0.1752	0.4747	1.752
3-13	0.3750	0.3875	0.17934	0.06265	0.23075	0.22251	0.25794	0.2229	0.5798	1.986
3-14	0.3750	0.3875	0.17832	0.06199	0.22926	0.22108	0.25634	0.2230	0.5801	1.981
3-15	0.3750	0.3875	0.17552	0.08540	0.24593	0.23462	0.27271	0.2231	0.5687	1.965
3-16	0.3750	0.3875	0.17425	0.08463	0.24414	0.23291	0.27078	0.2231	0.5690	1.958
3-17	0.3748	0.3878	0.15432	0.25377	0.36893	0.33428	0.39333	0.2244	0.5076	1.849
3-18	0.3748	0.3878	0.15224	0.25295	0.36668	0.33204	0.39083	0.2244	0.5077	1.837

TABLE 4.3-1.(3)

No.	F''_{GLw}	F_{GVi}	$\dfrac{F'_{GVi}}{10^{-2}}$	$\dfrac{F''_{GVi}}{10^{-2}}$	$-\Phi'_{Gi}$	$-\Theta'_{GVi}$	$-\Theta'_{GLw}$	$-\Theta'_{GLi}$
1- 1	0.1509	0.1999	1.1390	16.19	0.6060	0.7427	6.6673	6.6650
1- 2	0.1509	0.1999	1.1390	16.19	0.6060	0.7327	6.6673	6.6650
1- 3	0.1515	0.2009	1.1480	26.98	0.6581	0.7949	6.6673	6.6650
1- 4	0.2988	1.8820	4.4670	-17.05	3.7992	5.2548	3.3421	3.3094
1- 5	0.2988	1.8820	4.4670	-17.04	3.7991	5.1189	3.3421	3.3094
1- 6	0.2989	1.9100	4.4710	-14.37	3.8561	5.3333	3.3423	3.3087
1- 7	0.1513	0.2220	1.1440	24.78	0.5670	0.8111	6.6680	6.6631
1- 8	0.2500	1.1520	3.1270	4.157	1.9384	3.1056	4.0097	3.9734
1- 9	0.2007	0.5492	2.0140	15.54	0.8572	1.5073	5.0065	4.9823
1-10	0.2012	0.5550	2.0230	24.97	0.8812	1.5409	5.0065	4.9821
1-11	0.2498	1.1640	3.1220	0.2449	1.6921	3.0216	4.0148	3.9594
1-12	0.2498	1.1640	3.1220	0.2395	1.6921	3.0592	4.0148	3.9594
1-13	0.2501	1.1970	3.1280	5.415	1.7417	3.1475	4.0155	3.9575
2- 1	0.1604	0.0441	1.2870	1.502	0.1812	0.2518	6.2516	6.2457
2- 2	0.1604	0.0441	1.2870	1.502	0.1812	0.2531	6.2516	6.2457
2- 3	0.1665	0.0467	1.3840	20.92	0.3204	0.4332	6.2516	6.2455
2- 4	0.2798	0.2376	3.9160	0.2444	0.4290	0.6572	3.5799	3.5482
2- 5	0.2798	0.2376	3.9160	0.2418	0.4290	0.6658	3.5799	3.5482
2- 6	0.2849	0.2462	4.0590	17.03	0.5152	0.7670	3.5801	3.5477
2- 7	0.1406	0.0283	0.9886	1.906	0.1632	0.2228	7.1441	7.1394
2- 8	0.1406	0.0283	0.9886	1.906	0.1632	0.2218	7.1441	7.1394
2- 9	0.2208	0.1097	2.4380	2.692	0.2680	0.3918	4.5504	4.5319
2-10	0.1504	0.0331	1.1310	1.107	0.1451	0.2142	6.6684	6.6620
2-11	0.1574	0.0356	1.2360	21.78	0.2876	0.4107	6.6685	6.6617
2-12	0.2202	0.1045	2.4260	0.9501	0.2255	0.3637	4.5509	4.5306
2-13	0.2202	0.1045	2.4260	0.9507	0.2255	0.3623	4.5509	4.5306
2-14	0.2269	0.1103	2.5720	20.65	0.3437	0.5230	4.5511	4.5300
2-15	0.2008	0.0567	2.0160	1.789	0.1971	0.2955	5.0038	4.9897
2-16	0.2997	0.1916	4.4940	0.2048	0.3470	0.5756	3.3461	3.2983
2-17	0.2508	0.1171	3.1460	2.126	0.2902	0.4322	4.0065	3.9820
2-18	0.3481	0.3231	6.0660	-2.395	0.5404	0.8635	2.8752	2.8078
3- 1	0.1606	0.2583	1.2890	11.34	0.5179	0.8678	6.2508	6.2479
3- 2	0.2010	0.5124	2.0210	20.93	0.9087	1.3138	5.0019	4.9949
3- 3	0.2010	0.5128	2.0210	20.80	0.9091	1.6226	5.0019	4.9949
3- 4	0.1409	0.1673	0.9919	15.58	0.4385	0.6681	7.1434	7.1413
3- 5	0.1409	0.1673	0.9920	15.61	0.4389	0.6022	7.1434	7.1413
3- 6	0.1415	0.1687	1.0010	26.80	0.4784	0.7179	7.1434	7.1414
3- 7	0.1415	0.1687	1.0010	27.15	0.4807	0.6619	7.1434	7.1414
3- 8	0.1612	0.2532	1.2990	22.19	0.5710	0.9141	6.2509	6.2474
3- 9	0.1612	0.2531	1.2990	22.29	0.5715	0.7723	6.2509	6.2474
3-10	0.1617	0.2528	1.3070	.30.85	0.5926	0.9345	6.2509	6.2475
3-11	0.1617	0.2527	1.3070	31.41	0.5953	0.8111	6.2509	6.2476
3-12	0.1211	0.1018	0.7328	18.60	0.3910	0.5404	8.3337	8.3322
3-13	0.1313	0.1339	0.8617	23.26	0.4476	0.6398	7.6929	7.6907
3-14	0.1313	0.1338	0.8618	23.35	0.4483	0.5475	7.6929	7.6907
3-15	0.1313	0.1335	0.8625	24.31	0.4507	0.6424	7.6929	7.6907
3-16	0.1314	0.1333	0.8627	24.44	0.4517	0.5520	7.6929	7.6907
3-17	0.1318	0.1319	0.8688	32.43	0.4747	0.6661	7.6929	7.6908
3-18	0.1319	0.1317	0.8692	32.89	0.4775	0.5855	7.6929	7.6908

TABLE 4.3-1.(4)

No.	$\dfrac{q_{wx}x^{1/4}}{10^3}$ $(\mathrm{Wm^{-7/4}})$	$\dfrac{(q_{wx}x^{1/4})_{cal}}{10^3}$ $(\mathrm{Wm^{-7/4}})$	$\dfrac{q_{cx}x^{1/4}}{10^3}$ $(\mathrm{Wm^{-7/4}})$	$\dfrac{(q_{cx}x^{1/4})_{cal}}{10^3}$ $(\mathrm{Wm^{-7/4}})$	$\dfrac{\dot{m}_x x^{1/4}}{10^{-3}}$ $(\mathrm{kgm^{-7/4}s^{-1}})$	$\dfrac{(\dot{m}_x x^{1/4})_{cal}}{10^{-3}}$ $(\mathrm{kgm^{-7/4}s^{-1}})$
1- 1	4.7686	4.6840	0.0213	0.0198	2.1658	2.1288
1- 2	4.7683	4.6834	0.0210	0.0195	2.1658	2.1286
1- 3	5.3890	5.3050	0.6250	0.5852	2.1645	2.1453
1- 4	44.7361	44.3297	0.3893	0.3897	21.1517	21.1845
1- 5	44.7246	44.3168	0.3792	0.3740	21.1510	21.1858
1- 6	49.7320	49.2856	4.6714	4.6238	21.4646	21.5305
1- 7	5.2154	5.1619	0.0620	0.0588	2.6670	2.6436
1- 8	24.7385	24.5911	0.3077	0.3009	13.9987	14.0553
1- 9	10.3545	10.3018	0.1066	0.1027	7.0579	7.0630
1-10	11.9163	11.8699	1.5190	1.4731	7.1320	7.1676
1-11	21.8582	21.7090	0.2171	0.2142	14.9736	15.0892
1-12	21.8614	21.7124	0.2198	0.2178	14.9740	15.0890
1-13	25.4217	25.2330	3.1090	3.0731	15.3958	15.5472
2- 1	0.4800	0.4739	0.0031	0.0031	3.1101	3.0734
2- 2	0.4801	0.4740	0.0031	0.0031	3.1101	3.0735
2- 3	1.4161	1.4256	0.9087	0.9330	3.2929	3.2046
2- 4	2.5856	2.5947	0.0225	0.0218	16.7857	16.9955
2- 5	2.5859	2.5951	0.0228	0.0222	16.7857	16.9955
2- 6	4.3268	4.3791	1.6467	1.7448	17.3920	17.3223
2- 7	0.2872	0.2826	0.0050	0.0049	2.1011	2.0684
2- 8	0.2872	0.2826	0.0049	0.0049	2.1011	2.0686
2- 9	1.1064	1.1110	0.0201	0.0206	8.1881	8.2513
2-10	0.3190	0.3129	0.0034	0.0033	2.5892	2.5426
2-11	1.4431	1.4524	1.1016	1.1256	2.7815	2.6722
2-12	1.0050	1.0008	0.0103	0.0104	8.1915	8.1933
2-13	1.0050	1.0007	0.0103	0.0103	8.1915	8.1933
2-14	2.4706	2.5036	1.4115	1.4924	8.6304	8.3235
2-15	0.6722	0.6605	0.0135	0.0133	5.5601	5.4797
2-16	2.2745	2.2598	0.0454	0.0447	18.8690	19.0277
2-17	1.4712	1.4623	0.0267	0.0265	10.9053	10.9081
2-18	4.0932	4.0655	0.0802	0.0771	30.1799	30.7215
3- 1	6.3617	6.2532	0.0333	0.0315	2.7836	2.7379
3- 2	14.1755	13.9398	0.2132	0.2117	6.0330	5.9400
3- 3	14.2642	14.0951	0.2675	0.2768	6.0457	5.9771
3- 4	4.4737	4.4286	0.0445	0.0434	1.9358	1.9171
3- 5	4.4686	4.4159	0.0400	0.0386	1.9356	1.9137
3- 6	5.0068	5.0572	0.6070	0.5994	1.9269	1.9530
3- 7	4.9570	4.9668	0.5590	0.5450	1.9263	1.9373
3- 8	7.1930	7.1273	0.1244	0.1258	3.0601	3.0328
3- 9	7.1661	7.0737	0.1042	0.1031	3.0576	3.0198
3-10	7.7827	7.8475	0.8601	0.8740	3.0079	3.0318
3-11	7.6509	7.6186	0.7438	0.7363	3.0022	2.9932
3-12	3.1362	3.1153	0.0656	0.0647	1.3242	1.3158
3-13	4.3092	4.2852	0.1224	0.1228	1.7894	1.7795
3-14	4.2821	4.2376	0.1036	0.1023	1.7864	1.7684
3-15	4.3308	4.3118	0.1720	0.1726	1.7791	1.7713
3-16	4.2956	4.2505	0.1465	0.1444	1.7755	1.7576
3-17	4.6919	4.6999	0.6565	0.6578	1.7344	1.7378
3-18	4.5991	4.5539	0.5750	0.5650	1.7302	1.7156

surface in the case of the saturated vapors, while q_{cx} increases to maximum 13 percent of q_{wx} in the case of the air-water mixture of 100 K of superheat. For the latter case, the convective heat transfer in the vapor side cannot be neglected.

5. The effect of the diffusion term [the third term on the left-hand side of Eq. (4.1-8)] appears only in the values of $-\Theta_{GVi}$ and q_{cx} for the air-water mixture. These values are maximum 18 percent larger than those for the case where the diffusion term is neglected. However, the effect of the term upon q_{wx} and $\dot{m}_x$ is small because $q_{wx} \gg q_{cx}$.

4.3.2 DISTRIBUTIONS OF VELOCITY, TEMPERATURE, AND CONCENTRATION

Figures 4.3-1(a)–(e) show the distributions of F'_{GL}, F'_{GV} (solid lines), $\Theta_G = (T - T_w)/(T_{V\infty} - T_w)$ (dot-dash line), and Φ_G (broken line), where (a), (b), (c), (d), and (e) correspond to Nos. 1-1, 1-4, 2-15, 3-13, and 3-15 in Table 4.3-1, respectively.

From these figures we can find the following characteristics:

1. The distributions of temperature and velocity in the condensate film are nearly linear and parabolic [cf. Eq. (4.2-1)], respectively.

2. The velocity distribution in the vapor boundary layer resembles that of single phase free convection. When η_{GLi} or condensation mass flux increases, however, the vapor is drawn by the condensate and the maximum of velocity appears in the condensate film as seen in Fig. 4.3-1(b) for an ethanol-water mixture. The maximum value of velocity increases as the degree of superheat increases [as seen in the comparison between Figs. 4.3-1(d) and (e)].

3. The thickness of the vapor boundary layer for large η_{GLi} is thinner than that for small η_{GLi}. The magnitude of this difference is more remarkable than that for forced-convection condensation as seen in the comparison of the difference between Figs. 4.3-1(a) and (b) with the difference between Figs. 3.3-1(a) and (b).

4. The dimensionless vapor boundary layer thickness is large in the case of small values of M_1/M_2 and R [as seen in the comparison between Figs. 4.3-1(a), (b) and Fig. 4.3-1(c)].

5. The temperature and concentration in the vapor boundary layer vary monotonously, and the thickness of the former is thinner than that of the latter, which stems from the fact that $Pr_V > Sc$.

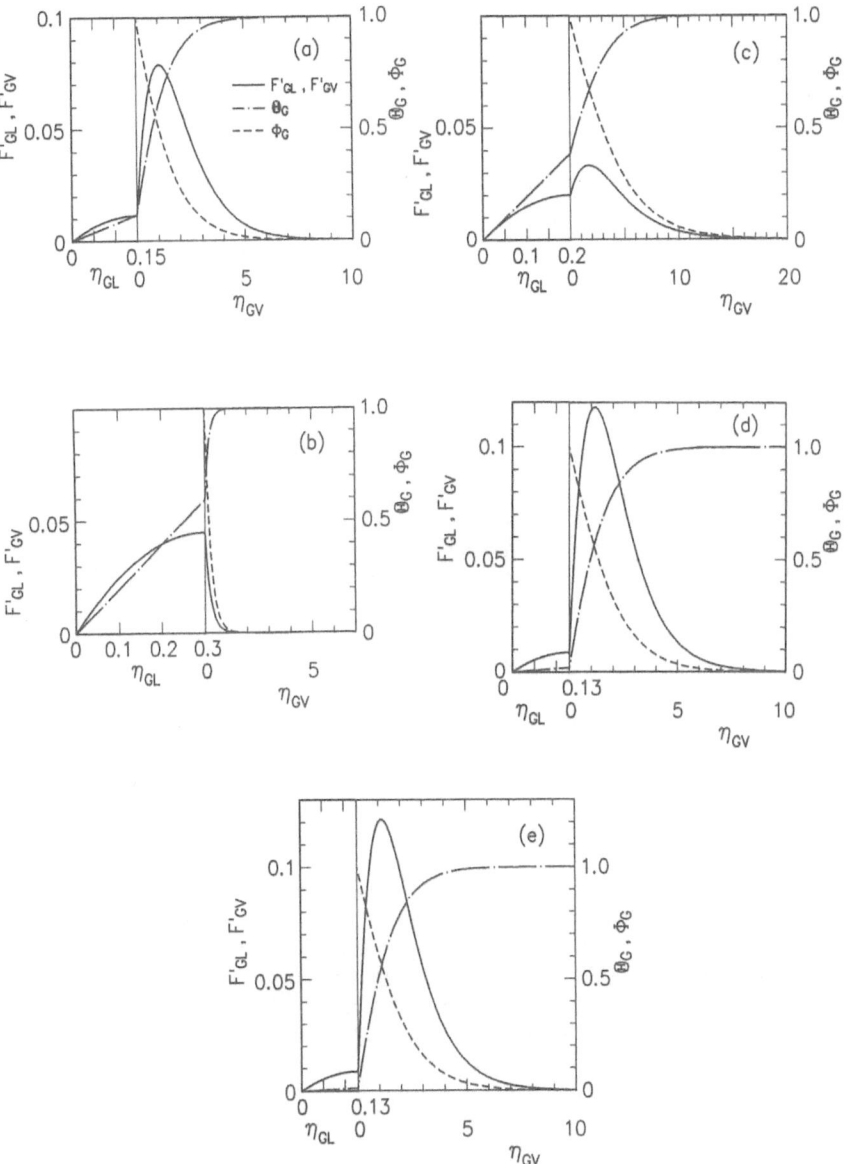

FIGURE 4.3-1. Distributions of velocities F'_{GL}, F'_{GV}, temperature Θ_G, and mass concentration Φ_G across the condensate film and the vapor boundary layer in the case of free-convection condensation (No. corresponds to that in Table 4.3-1). (a) No. 1-1: ethanol-water, $M_1/M_2 = 1.779$, $Sc = 0.73$, $Pr_V = 0.92$, $R = 176$. (b) No. 1-4: ethanol-water, $Sc = 0.67$, $Pr_V = 0.90$, $R = 210$. (c) No. 2-15: CFC114-CFC11, $M_1/M_2 = 1.244$, $Sc = 0.44$, $Pr_V = 0.79$, $R = 21$. (d) No. 3-13: air-water, $M_1/M_2 = 1.608$, $Sc = 0.55$, $Pr_V = 0.73$, $R = 180$. (e) No. 3-15: air-water, $\Delta T_{V\infty} = 10$ K, $Sc = 0.55$, $Pr_V = 0.73$, $R = 180$.

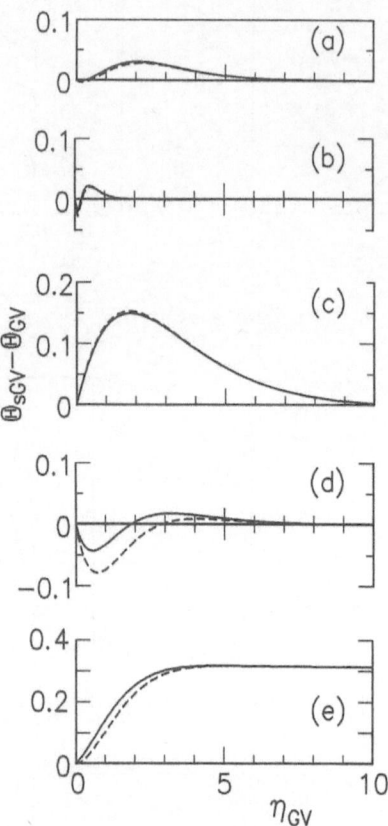

FIGURE 4.3-2. Distributions of $\Theta_{sGV} - \Theta_{GV}$ across the vapor boundary layer in the case of free-convection condensation (No. corresponds to that in Table 4.3-1). (a) No. 1-1,2: ethanol-water, $\eta_{GLi} = 0.15$, $T_{V\infty} = 98°C$, $W_{1V\infty} = 0.145$, $W_{1Vi} = 0.382$. (b) No. 1-4,5: ethanol-water, $\eta_{GLi} = 0.30$, $T_{V\infty} = 98°C$, $W_{1V\infty} = 0.145$, $W_{1Vi} = 0.596$. (c) No. 2-4,5: CFC114-CFC11r, $\eta_{GLi} = 0.28$, $T_{V\infty} = 75°C$, $W_{1V\infty} = 0.219$, $W_{1Vi} = 0.309$. (d) No. 1-13,14: air-water, $\eta_{GLi} = 0.13$, $T_{V\infty} = 90°C$, $W_{1V\infty} = 0.407$, $W_{1Vi} = 0.806$. (e) No. 1-15,16: air-water, $\eta_{GLi} = 0.13$, $T_{V\infty} = 90°C$, $W_{1V\infty} = 0.407$, $W_{1Vi} = 0.797$, $\Delta T_{V\infty} = 10$ K.

4.3.3   POSSIBILITY OF THE APPEARANCE OF SUBCOOLING IN THE VAPOR BOUNDARY LAYER

Figures 4.3-2(a)–(e) show the distributions of the difference between the vapor temperature and the saturation temperature corresponding to concentration. The coordinates and lines used in the figures are the same as in Figs. 3.3-2(a)–(g). Characteristics of the results are also similar to the case of forced-convection condensation.

In the comparison between Figs. 4.3-2(a) and (b) for ethanol-water mixtures, the degree of subcooling is larger for larger η_{GLi}. In the case of a CFC114-CFC11 mixture, subcooling does not appear [as seen in Fig. 4.3-2(c)]. In Fig. 4.3-2(d), for an air-water mixture, the degree of subcooling and the effect of the diffusion term upon it are larger in comparison with the cases of the other mixtures. However, the subcooling disappears with some superheating bulk vapor mixture [as seen in Fig. 4.3-2(e)].

4.4   Formulas of the Boundary Values Θ'_{GLw}, Θ'_{GLi}, and Θ'_{GVi} for Dimensionless Temperature

The local Nusselt numbers of Nu_{Lwx}, Nu_{Lix}, and Nu_{cx}, which are defined by Eqs. (3.4-4), (3.4-5), and (3.4-6), respectively, are expressed as

$$Nu_{Lwx} = \frac{1}{\sqrt{2}}(-\Theta'_{GLw})Ga_x^{\frac{1}{4}}, \tag{4.4-1}$$

$$Nu_{Lix} = \frac{1}{\sqrt{2}}(-\Theta'_{GLi})Ga_x^{\frac{1}{4}}, \tag{4.4-2}$$

$$Nu_{cx} = \frac{1}{\sqrt{2}}(-\Theta'_{GVi})\left(\frac{\nu_L}{\nu_V}\right)^{\frac{1}{2}} Ga_x^{\frac{1}{4}}. \tag{4.4-3}$$

As for the heat transfer in the condensate film, Figs. 4.4-1(a) and (b) show the relations of $-\Theta'_{GLw}$ versus $\dot{M}_{GL}$ and $-\Theta'_{GLi}$ versus $\dot{M}_{GL}$, respectively. The solid lines in these figures represent the equation, which was obtained by Fujii et al.:[1]

$$-\Theta'_{GLw} \approx -\Theta'_{GLi} = \dot{M}_{GL}^{-\frac{1}{3}}. \tag{4.4-4}$$

This equation can correlate the data within an error of 0.8 percent for $-\Theta'_{GLw}$ and 1.5 percent for $-\Theta'_{GLi}$.

We can derive the Nu_{cx} expression by referring to the characteristics of single phase heat and mass transfer as in the case of forced-convection condensation. Tanaka[2] proposed the following equation for heat transfer in free-convection heat and mass transfer of a binary vapor mixture on a

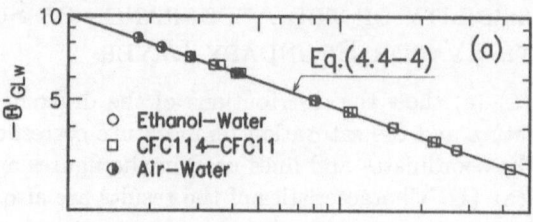

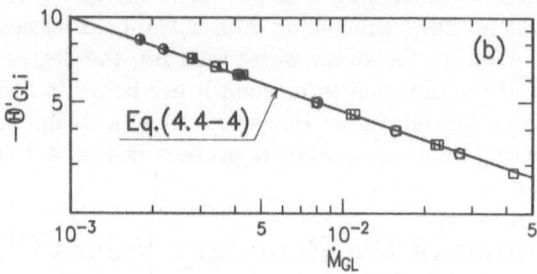

FIGURE 4.4-1. Correlation of (a) $-\Theta'_{GLw}$ versus $\dot{M}_{GL}$ and (b) $-\Theta'_{GLi}$ versus $\dot{M}_{GL}$.

vertical surface with negligibly small suction velocity,

$$
\begin{aligned}
Nu_{cx} &= C_G(Pr_V)\left\{\left(\frac{g\chi_i^+ x^3}{\nu_V^2}\right)Pr_V\right\}^{\frac{1}{4}} \\
&= C_G(Pr_V)\left(\frac{\chi_i^+}{\omega_i}Gr_x Pr_V\right)^{\frac{1}{4}},
\end{aligned} \tag{4.4-5}
$$

where

$$
C_G(Pr_V) = \frac{3}{4}\left(\frac{Pr_V}{2.4 + 4.9\sqrt{Pr_V} + 5Pr_V}\right)^{\frac{1}{4}} \tag{4.4-6a}
$$

$$
\approx 0.4 Pr_V^{0.1} \qquad (0.4 \lesssim Pr_V \lesssim 2) \tag{4.4-6b}
$$

$$
\approx 0.415 Pr_V^{0.05} \qquad (\ 2 \lesssim Pr_V \lesssim 15), \tag{4.4-6c}
$$

$$
Gr_x = \frac{g\omega_i x^3}{\nu_V^2}, \tag{4.4-7}
$$

$$
\omega_i = \omega_W + \omega_T - \omega_W\omega_T, \tag{4.4-8}
$$

$$
\chi_i^+ = \omega_W\left(\frac{Pr_V}{Sc}\right)^{\frac{1}{2}} + \omega_T - \omega_W\omega_T, \tag{4.4-9}
$$

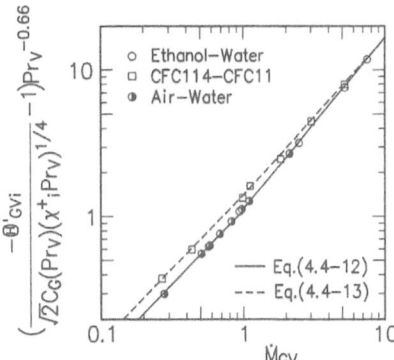

FIGURE 4.4-2. Correlation of $[\{-\Theta'_{GVi}/\sqrt{2}C_G(Pr_V)(\chi_i^+ Pr_V)^{1/4}\} - 1]Pr_V^{-0.66}$ versus $\dot{M}_{GV}$.

and Eq. (4.4-6a) was proposed by LeFevre.[3] We can define the dimensionless condensation mass flux by using physical properties of vapor mixture as

$$\dot{M}_{GV} = \frac{\dot{m}_x x}{\mu_V}\left(\frac{Gr_x}{4}\right)^{-\frac{1}{4}} \qquad (4.4\text{-}10a)$$

$$= \omega_i^{-\frac{1}{4}} R\dot{M}_{GL}. \qquad (4.4\text{-}10b)$$

Since Eq. (4.4-3) approaches Eq. (4.4-5) in the case of $\dot{m}_x \to 0$,

$$\frac{-\Theta'_{GVi}}{\sqrt{2}C_G(Pr_V)(\chi_i^+ Pr_V)^{\frac{1}{4}}} \to 1. \qquad (4.4\text{-}11)$$

Figure 4.4-2 shows the relation of $[\{-\Theta'_{GVi}/\sqrt{2}C_G(Pr_V)(\chi_i^+ Pr_V)^{1/4}\} - 1]Pr_V^{-0.66}$ versus $\dot{M}_{GV}$ for the data in the case of $c_{p12} = 0$ for the mixtures of ethanol-water, air-water, and CFC114-CFC11. From this figure we can obtain the following equation:

$$\frac{-\Theta'_{GVi}}{\sqrt{2}C_G(Pr_V)(\chi_i^+ Pr_V)^{\frac{1}{4}}} = 1 + kPr_V^{0.66}\dot{M}_{GV}^\ell, \qquad (4.4\text{-}12)$$

where for the mixtures of ethanol-water and air-water,

$$k = 1.13 \quad \text{and} \quad \ell = 1.17 \quad \text{for} \quad \dot{M}_{GV} \gtrsim 1, \qquad (4.4\text{-}13a)$$
$$k = 1.1 \quad \text{and} \quad \ell = 1 \quad \text{for} \quad \dot{M}_{GV} \lesssim 1; \qquad (4.4\text{-}13b)$$

for the CFC114-CFC11 mixture,

$$k = 1.4 \quad \text{and} \quad \ell = 1.07 \quad \text{for} \quad \dot{M}_{GV} \gtrsim 1, \qquad (4.4\text{-}14a)$$
$$k = 1.4 \quad \text{and} \quad \ell = 1 \quad \text{for} \quad \dot{M}_{GV} \lesssim 1. \qquad (4.4\text{-}14b)$$

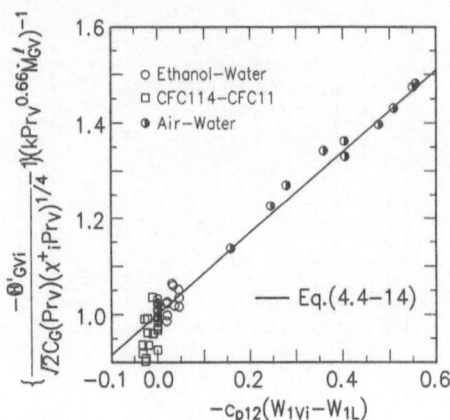

FIGURE 4.4-3. The effect of $c_{p12}(W_{1Vi} - W_{1L})$ upon Θ'_{GVi}.

Figure 4.4-3 shows the relation of $[\{-\Theta'_{GVi}/\sqrt{2}C_G(Pr_V)(\chi_i^+ Pr_V)^{1/4}\} - 1](kPr_V^{0.66}\dot{M}_{GV}^\ell)^{-1}$ versus $-c_{p12}(W_{1Vi} - W_{1L})$. From this figure we can obtain the equation

$$\frac{-\Theta'_{GVi}}{\sqrt{2}C_G(Pr_V)(\chi_i^+ Pr_V)^{\frac{1}{4}}} = 1 + kPr_V^{0.66}\dot{M}_{GV}^\ell\{1 - 0.85c_{p12}(W_{1Vi} - W_{1L})\}.$$

(4.4-15)

4.5   Formula of the Boundary Value Φ'_{Gi} for Normalized Concentration[4]

The local Sherwood number Sh_x, which is defined by Eq. (3.5-2), is expressed as

$$Sh_x = \frac{1}{\sqrt{2}}(-\Phi'_{Gi})\left(\frac{\nu_L}{\nu_V}\right)^{\frac{1}{2}} Ga_x^{\frac{1}{4}}.$$

(4.5-1)

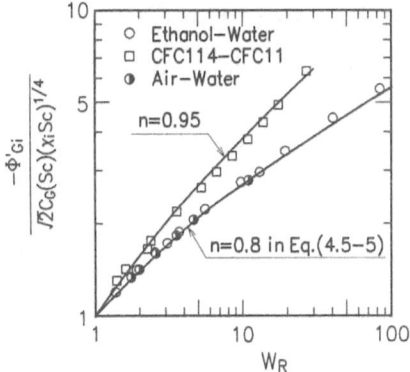

FIGURE 4.5-1. Correlation of $-\Phi'_{Gi}/\sqrt{2}C_G(Sc)(\chi_i Sc)^{1/4}$ versus W_R.

Tanaka[2] also proposed the equation for mass transfer in free-convection heat and mass transfer of a binary vapor mixture as

$$Sh_x = C_G(Sc)\left(\frac{\chi_i}{\omega_i} Gr_x Sc\right)^{\frac{1}{4}}, \qquad (4.5\text{-}2)$$

where $C_G(Sc)$ is the function of Sc with the same form as Eq. (4.4-6) and

$$\chi_i = \omega_W + \omega_T \left(\frac{Sc}{Pr_V}\right)^{\frac{1}{2}} - \omega_W \omega_T. \qquad (4.5\text{-}3)$$

When the condensation mass flux $\dot{m}_x$ approaches zero, it can be postulated that Eq. (4.5-1) approaches Eq. (4.5-2), and consequently

$$\frac{-\Phi'_{Gi}}{\sqrt{2}C_G(Sc)(\chi_i Sc)^{\frac{1}{4}}} \to 1. \qquad (4.5\text{-}4)$$

Figure 4.5-1 shows the relation of $-\Phi'_{Gi}/\sqrt{2}C_G(Sc)(\chi_i Sc)^{1/4}$ versus W_R, plotted from the data in Table 4.3-1. The lines in the figure represent the equation

$$\frac{-\Phi'_{Gi}}{\sqrt{2}C_G(Sc)(\chi_i Sc)^{\frac{1}{4}}} = \left(\frac{2}{1+W_R}\right)^{0.5} W_R{}^n, \quad 1 \le W_R < 100, \quad (4.5\text{-}5)$$

where
$$\begin{array}{ll} n = 0.8 & \text{for ethanol-water and air-water} \\ n = 0.95 & \text{for CFC114-CFC11} \end{array} \Bigg\}. \qquad (4.5\text{-}6)$$

Equation (4.5-5) for $n = 0.8$ was obtained for an air-water mixture by Fujii et al.,[1] and that for $n = 0.95$ will be applicable to the case of small values of Sc, R, and χ_i.

4.6   Algebraic Method for Calculating Condensation Mass Flux and Heat Flux

The substitution of Eqs. (4.4-4) and (4.4-12) into Eq. (4.1-24) yields

$$
\dot{M}_{GL}^{-\frac{1}{3}} = \underbrace{\frac{Pr_L \dot{M}_{GL}}{H_i}}_{②} + \underbrace{\sqrt{2}C_G (Pr_V)\frac{\lambda_V}{\lambda_L}\left(\frac{\nu_L}{\nu_V}\right)^{\frac{1}{2}}\frac{(T_{V\infty}-T_i)}{(T_i - T_w)}(\chi_i^+ Pr_V)^{\frac{1}{4}}}_{③}
$$
$$
\times \underbrace{\left[1 + kPr_V^{0.66}\left(\omega_i^{-\frac{1}{4}}R\dot{M}_{GL}\right)^{\ell}\{1 - 0.85c_{p12}(W_{1Vi} - W_{1L})\}\right]}_{④},
$$

①——   ②——   ③——

$$(4.6\text{-}1)$$

where term ① is the dimensionless heat flux through the condensate film, term ② the dimensionless latent heat released at the vapor-liquid interface, term ③ the dimensionless heat flux of the vapor phase side, which is negligibly small for saturated vapors, and term ④ the effect of the diffusion term, which is relatively large for an air-water mixture.

The substitution of Eqs. (4.1-21) and (4.5-5) into Eq. (4.1-31) yields

$$
\dot{M}_{GL} = \frac{2C_G(Sc)(\chi_i Sc)^{\frac{1}{4}}}{ScR}\frac{(W_R - 1)}{(W_R + 1)^{0.5}W_R^{1-n}}.
$$
$$(4.6\text{-}2)$$

The algebraic equations consist of Eqs. (4.6-1) and (4.6-2), the applicable ranges of which are estimated as follows: $Pr_L = 1\text{-}10$, $R = 20\text{-}250$, $Pr_V = 0.7\text{-}1$, $Sc = 0.4\text{-}0.8$, $\Delta T_{V\infty} \leq 100$ K, and $W_R = 1\text{-}100$. The case of $W_R > 100$ appears when $W_{1L} \approx W_{1V\infty}$, which means total condensation. In this case, therefore, the value of T_i is determined from only the phase equilibrium diagram without Eq. (4.6-2).

All relevant physical quantities can be obtained from respective definition equations and the algebraic solution, e.g.,

$$
\dot{m}_x = \left(\frac{\mu_L \dot{M}_{GL}}{x}\right)\left(\frac{Ga_x}{4}\right)^{\frac{1}{4}},
$$
$$(4.6\text{-}3)$$

$$
q_{wx} = \frac{\lambda_L(T_i - T_w)\dot{M}_{GL}^{-\frac{1}{3}}}{x}\left(\frac{Ga_x}{4}\right)^{\frac{1}{4}}.
$$
$$(4.6\text{-}4)$$

The values of $\dot{m}_x$ and q_{wx} obtained by the algebraic solution agree with those of the similarity solution within an error of two percent.

The condensation mass flux and wall heat flux averaged over the length $x = 0$ to ℓ are expressed as

$$\bar{m} = \frac{4}{3}(\dot{m}_x)_{x=\ell} ,$$ (4.6-5)

$$\bar{q}_w = \frac{4}{3}(q_{wx})_{x=\ell} .$$ (4.6-6)

4.7 Relations between Relevant Physical Quantities and Dimensionless Functions for Free-Convection Condensation

For the condensate film :

$$c_L = \left(\frac{g}{4\nu_L^2}\right)^{\frac{1}{4}} ,$$ (4.7-1)

$$U_L = 4\nu_L c_L^2 x^{\frac{1}{2}} F_{GL}' ,$$ (4.7-2)

$$V_L = -\nu_L c_L x^{-\frac{1}{4}} (3F_{GL} - \eta_{GL} F_{GL}') ,$$ (4.7-3)

$$\frac{\partial U_L}{\partial x} = \nu_L c_L^2 x^{-\frac{1}{2}} (2F_{GL}' - \eta_{GL} F_{GL}'') ,$$ (4.7-4)

$$\frac{\partial U_L}{\partial y} = 4\nu_L c_L^3 x^{\frac{1}{4}} F_{GL}'' ,$$ (4.7-5)

$$\frac{\partial^2 U_L}{\partial y^2} = 4\nu_L c_L^4 F_{GL}''' ,$$ (4.7-6)

$$\frac{\partial T_L}{\partial x} = -\frac{1}{4}(T_i - T_w)x^{-1}\eta_{GL}(-\Theta_{GL}') ,$$ (4.7-7)

$$\frac{\partial T_L}{\partial y} = (T_i - T_w)c_L x^{-\frac{1}{4}}(-\Theta_{GL}') ,$$ (4.7-8)

$$\frac{\partial^2 T_L}{\partial y^2} = (T_i - T_w)c_L^2 x^{-\frac{1}{2}}(-\Theta_{GL}'') ,$$ (4.7-9)

$$\delta = c_L^{-1} x^{\frac{1}{4}} \eta_{GLi} ,$$ (4.7-10)

$$\frac{d\delta}{dx} = \frac{1}{4}c_L^{-1} x^{-\frac{3}{4}} \eta_{GLi} ,$$ (4.7-11)

$$\dot{m}_x = 3\mu_L c_L x^{-\frac{1}{4}} F_{GLi} .$$ (4.7-12)

For the vapor boundary layer :

$$c_V = \left(\frac{g}{4\nu_V^2}\right)^{\frac{1}{4}} ,$$ (4.7-13)

$$U_V = 4\nu_V c_V^2 x^{\frac{1}{2}} F'_{GV}, \tag{4.7-14}$$

$$V_V = -3\nu_V c_V x^{-\frac{1}{4}} F_{GV} + \left(4\nu_V c_V^2 x^{\frac{1}{2}} \frac{d\delta}{dx} + \nu_V c_V x^{-\frac{1}{4}} \eta_{GV}\right) F'_{GV}, \tag{4.7-15}$$

$$\frac{\partial U_V}{\partial x} = 2\nu_V c_V^2 x^{-\frac{1}{2}} F'_{GV} - 4\nu_V c_V^3 x^{\frac{1}{4}} \frac{d\delta}{dx} F''_{GV} - \nu_V c_V^2 x^{-\frac{1}{2}} \eta_{GV} F''_{GV}, \tag{4.7-16}$$

$$\frac{\partial U_V}{\partial y} = 4\nu_V c_V^3 x^{\frac{1}{4}} F''_{GV}, \tag{4.7-17}$$

$$\frac{\partial^2 U_V}{\partial y^2} = 4\nu_V c_V^4 F'''_{GV}, \tag{4.7-18}$$

$$\frac{\partial T_V}{\partial x} = (T_{V\infty} - T_i) c_V \frac{d\delta}{dx} x^{-\frac{1}{4}} \Theta'_{GV} + \frac{1}{4}(T_{V\infty} - T_i)\eta_{GV} x^{-1} \Theta'_{GV}, \tag{4.7-19}$$

$$\frac{\partial T_V}{\partial y} = (T_{V\infty} - T_i) c_V x^{-\frac{1}{4}} (-\Theta'_{GV}), \tag{4.7-20}$$

$$\frac{\partial^2 T_V}{\partial y^2} = (T_{V\infty} - T_i) c_V^2 x^{-\frac{1}{2}} (-\Theta''_{GV}), \tag{4.7-21}$$

$$\frac{\partial W_1}{\partial x} = (W_{1Vi} - W_{1V\infty})\left(c_V x^{-\frac{1}{4}} \frac{d\delta}{dx} + \frac{1}{4} x^{-1} \eta_{GV}\right)(-\Phi'_G), \tag{4.7-22}$$

$$\frac{\partial W_1}{\partial y} = -(W_{1Vi} - W_{1V\infty}) c_V x^{-\frac{1}{4}} (-\Phi'_G), \tag{4.7-23}$$

$$\frac{\partial^2 W_1}{\partial y^2} = -(W_{1Vi} - W_{1V\infty}) c_V^2 x^{-\frac{1}{2}} (-\Phi''_G), \tag{4.7-24}$$

$$\dot{m}_x = 3\mu_V c_V x^{-\frac{1}{4}} F_{GVi}. \tag{4.7-25}$$

REFERENCES

1. Fujii, T., H. Uehara, K. Mihara, and H. Takashima, Body Force Convection Condensation in the Presence of Noncondensables (in Japanese), *Reports of Research Institute of Industrial Science, Kyushu University*, No. 67, 23–41 (1978).

2. Tanaka, H., On Expressions for Local Nusselt Number and Local Sherwood Number Concerning Simultaneous Heat and Mass Transfer in Free Convection from a Vertical Plate (in Japanese), *Reports of Research Institute of Industrial Science, Kyushu University*, No. 78, 47–52 (1985).

3. LeFevre, E. J., Laminar Free Convection from a Vertical Plane Surface, *Proc. 9th Int. Congr. Appl. Mech., Brussels*, 4, 168–174 (1956).

4. Koyama, Sh., M. Watabe, and T. Fujii, The Gravity Controlled Film Condensation of Saturated and Superheated Binary Vapour Mixtures on a Vertical Plate (in Japanese), *Trans. Jpn. Soc. Mech. Eng.*, 52, 474, B, 827–834 (1986).

5

Condensation of Pure Vapors

Since condensation of pure vapors is a special case of binary vapor mixtures, any repeated discussion is unnecessary so far as we evaluate wall heat flux and condensation mass flux. Also, a comprehensive review on laminar film condensation of a saturated pure vapor was recently presented by Rose.[1] In this chapter, however, the relation between the present results and reported expressions, valuation on reported approximate solutions, and the effect of degree of superheat of vapor upon wall heat flux and condensation mass flux are discussed by using the equations obtained in Chapters 3 and 4. Furthermore, correlation equations for combined forced- and free-convection condensation and approximate solutions for the case of uniform wall heat flux are presented.

5.1 Forced-Convection Condensation of Saturated Pure Vapors

Equation (3.6-1) into which $T_{V\infty} = T_i = T_s$ are substituted, pertains to the case of forced-convection condensation of a saturated pure vapor, and hence the dimensionless mass flux $\dot{M}_{FV}$ can be solved from

$$0.433\left(1.367 - \frac{0.432}{\sqrt{2\dot{M}_{FV}}} + \frac{1}{2\dot{M}_{FV}}\right)^{\frac{1}{2}}(1 + 0.320H^{0.87})^{-1} = \frac{Pr_L}{RH}\dot{M}_{FV}\,,$$

$$(5.1\text{-}1)$$

where

$$H = \frac{c_{pL}(T_s - T_w)}{\Delta h_V}\,.$$

$$(5.1\text{-}2)$$

H is called the phase change number. The normal practical ranges of the parameters H, H/Pr_L, and RH/Pr_L are

$$H = 10^{-3} - 0.1\,, \qquad (5.1\text{-}3a)$$
$$H/Pr_L = 1 \times 10^{-4} - 5 \times 10^{-2}\,, \qquad (5.1\text{-}3b)$$
$$RH/Pr_L = 0.01 - 10\,. \qquad (5.1\text{-}3c)$$

The dimensionless heat transfer coefficient Nu_{Lwx} is derived from Eqs. (3.4-4) and (3.4-8) as

$$Nu_{Lwx} = 0.433 \left(1.367 - \frac{0.432}{\sqrt{2\dot{M}_{FV}}} + \frac{1}{2\dot{M}_{FV}} \right)^{\frac{1}{2}} Re_{Lx}^{\frac{1}{2}}. \qquad (5.1\text{-}4)$$

In the case where the condensate film is so thin that the effect of the convection term upon the temperature profile is negligibly small, Eq. (5.1-1) can be simplified as

$$0.433 \left(1.367 - \frac{0.432}{\sqrt{2\dot{M}_{FV}}} + \frac{1}{2\dot{M}_{FV}} \right)^{\frac{1}{2}} = \frac{Pr_L}{RH}\dot{M}_{FV} \qquad (5.1\text{-}5)$$

and the following equations are valid

$$q_{wx} = \dot{m}_x \Delta h_V, \qquad (5.1\text{-}6)$$

$$\dot{M}_{FV} = R\dot{M}_{FL} = \frac{R\dot{m}_x x}{\mu_L Re_{Lx}^{\frac{1}{2}}} = \left(\frac{RH}{Pr_L} \right) \frac{Nu_{Lwx}}{Re_{Lx}^{\frac{1}{2}}}. \qquad (5.1\text{-}7)$$

Solving $\dot{M}_{FV}$ from Eq. (5.1-5) with RH/Pr_L as a parameter and transforming $\dot{M}_{FV}$ to Nu_{Lwx} by using Eq. (5.1-7), we obtain the relation of $Nu_{Lwx}/Re_{Lx}^{1/2}$ versus RH/Pr_L as shown in Fig. 5.1-1. The curve in the figure agrees with the following equation of Fujii and Uehara[2] within an error of one percent in the range of $0.01 \lesssim RH/Pr_L \lesssim 5$:

$$Nu_{Lwx} = \chi Re_{Lx}^{\frac{1}{2}}, \qquad (5.1\text{-}8)$$

where

$$\chi = 0.45 \left(1.20 + \frac{Pr_L}{RH} \right)^{\frac{1}{3}}. \qquad (5.1\text{-}9)$$

The Nusselt number averaged over $0 \le x \le \ell$ is expressed as

$$\overline{Nu}_{Lw\ell} = 2\chi Re_{L\ell}^{\frac{1}{2}}. \qquad (5.1\text{-}10)$$

Equation (5.1-8) agrees with the approximate solution of Cess[3] and the numerical solution of Sparrow et al.[4] within an error of one percent in the range of $0.035 \lesssim RH/Pr_L \lesssim 5$. In the case where the effect of convection term is appreciable, we can obtain more accurate values of Nu_{Lwx} by solving $\dot{M}_{FV}$ from Eq. (5.1-1) and substituting it into Eq. (5.1-4).

Koh[5] graphically presented his numerical solutions for various combinations of three parameters of $R = 10$, 100, and 500, $Pr_L = 0.003$, 0.008, 0.03, 1, 10, and 100, and $0.002 \lesssim RH/Pr_L \lesssim 20$, and found the following

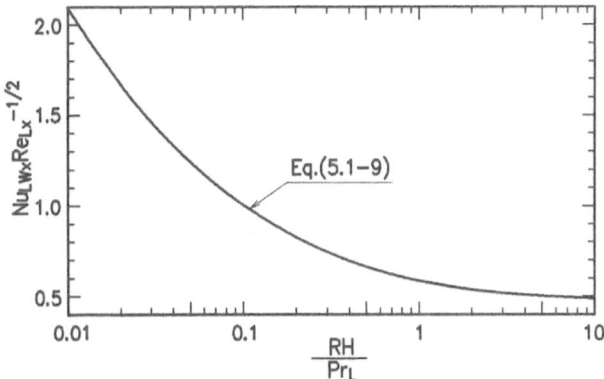

FIGURE 5.1-1.   Correlation of $Nu_{Lwx}Re_{Lx}^{-1/2}$ versus RH/Pr_L for the normal ranges of the variables of a pure vapor.

tendency in the solutions: the values of $Nu_{Lwx}Re_{Lx}^{-1/2}$ for $Pr_L = 10$ and 100 and those for $Pr_L = 1$ become, respectively, larger and smaller than those of Cess[3] when the value of RH/Pr_L is larger than a value depending on R and Pr_L. The amounts of the increase and decrease are remarkable for small values of R (note that there is a careless mistake in the graphical plotting). Now we discuss the case of large values of RH/Pr_L and Pr_L, though it is very rare in practice.

Fujii and Lee[6] obtained the similarity solution for $R = 5, 10, 100$ and $Pr_L = 1, 3, 5, 10, 20, 50, 100$. The relation of $Nu_{Lwx}Re_{Lx}^{-1/2}$ versus RH/Pr_L for $R = 10$ is shown in Fig. 5.1-2 as an example. The symbols $\triangle, \bigcirc$, and $\square$ represent the cases of $Pr_L = 1, 10$, and 100, respectively, of Koh's results,

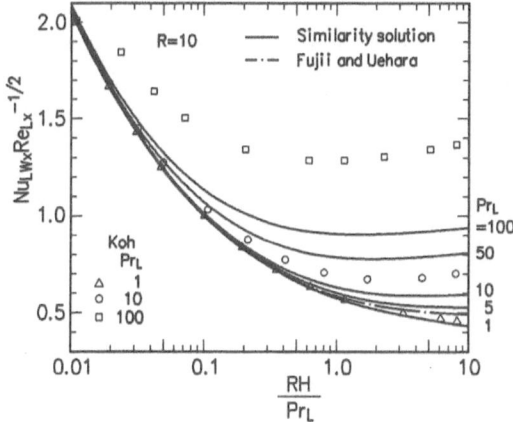

FIGURE 5.1-2.   Correlation of $Nu_{Lwx}Re_{Lx}^{-1/2}$ versus RH/Pr_L for the range of $Pr_L=1-100$ and $R=10$.

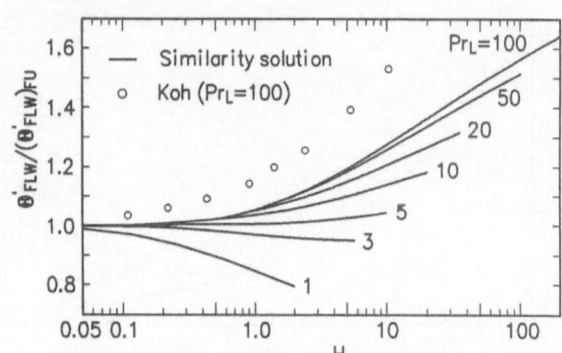

FIGURE 5.1-3. Correlation of $\Theta'_{FLw}/(\Theta'_{FLw})_{FU}$ or $(Nu_{Lwx})_{sim}/(Nu_{Lwx})_{FU}$ versus H for the range of $Pr_L = 1$–100.

which are quoted from Table 1 in Ref. 7. His results for $Pr_L = 10$ and 100 are much larger than those of Fujii and Lee.

Fujii and Lee[6] discovered the fact that the ratio of Θ'_{FLw} [proportional to $(Nu_{Lwx})_{sim}$] obtained from the similarity solution to $(\Theta'_{FLw})_{FU}$ [proportional to $(Nu_{Lwx})_{FU}$] from Eq. (5.1-8) is a function of H and Pr_L, almost independently of R, as shown in Fig. 5.1-3, i.e.,

$$\frac{\Theta'_{FLw}}{(\Theta'_{FLw})_{FU}} = \frac{(Nu_{Lwx})_{sim}}{(Nu_{Lwx})_{FU}} = C(H, Pr_L). \qquad (5.1-11)$$

Hence, $C(H, Pr_L)$ is a correction factor for Eq. (5.1-8) in the case of a large value of H.

Fujii and Lee[6] also found that the following relation between Θ'_{FLi} and Θ'_{FLw} is valid irrespective of Pr_L and R:

$$(-\Theta'_{FLi}) = (1 + 0.320H^{0.87})^{-1}(-\Theta'_{FLw}) \qquad (5.1-12)$$

as seen in Fig. 5.1-4. As a matter of fact, this relation has been used already in Eq. (3.4-10).

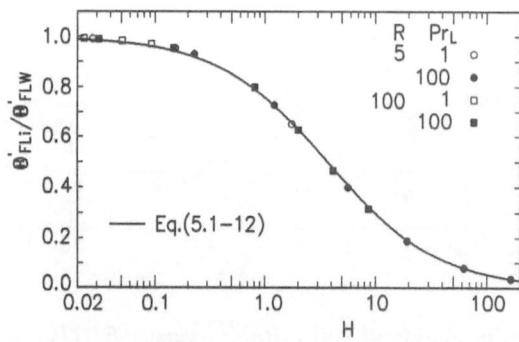

FIGURE 5.1-4. Correlation of $\Theta'_{FLi}/\Theta'_{FLw}$ or $(\dot{M}_{FV})_{sim}/(\dot{M}_{FV})_{FU}$ versus H.

The elimination of Θ'_{FLw} from Eqs. (5.1-11) and (5.1-12) gives the equation

$$\frac{-\Theta'_{FLi}}{(\Theta'_{FLw})_{FU}} = \frac{C(H, Pr_L)}{1 + 0.320H^{0.87}} = \frac{(\dot{M}_{FV})_{sim}}{(\dot{M}_{FV})_{FU}}, \qquad (5.1\text{-}13)$$

where $(\dot{M}_{FV})_{sim}$ is based on the relation $q_{ix} = \dot{m}_x \Delta h_V$ and $(\dot{M}_{FV})_{FU}$ is the value obtained from Eq. (5.1-5) or Eq. (5.1-8). We can find from the comparison between Figs. 5.1-3 and 5.1-4 that the difference between $(\dot{M}_{FV})_{sim}$ and $(\dot{M}_{FV})_{FU}$ is larger than the difference between $(Nu_{Lwx})_{sim}$ and $(Nu_{Lwx})_{FU}$, i.e., the effect of the convection term appears more remarkably in the condensation mass flux than in the wall heat flux.

In the case where the value of R is nearly equal to one, i.e., in the state near the critical point, the variation of all relevant physical properties with temperature should be taken into account as shown later in Chapter 10.

Figure 5.1-5 shows the relation between $Nu_{Lwx}Re_{Lx}^{-1/2}$ and H/Pr_L for $Pr_L = 0.003$–0.03 and $R = 1000$, corresponding to the case of metal vapors. In the figure, the symbol $\bigcirc$, the solid line, and the broken line correspond to Koh's result for $R = 500$ (referred from Table 1 in Ref. 7), Fujii and Lee's[6] result, and Rose's[8] formula, respectively. Fujii and Lee's results agree well with those of Koh, and can be correlated by the equation

$$\frac{Nu_{Lwx}}{\sqrt{Re_{Lx}}} = 0.5 - 0.13\left(\frac{H}{Pr_L}\right)^{0.85}$$

$$(0.01 \leq \frac{H}{Pr_L} \leq 1, \quad 0.003 \leq Pr_L \leq 0.03). \qquad (5.1\text{-}14)$$

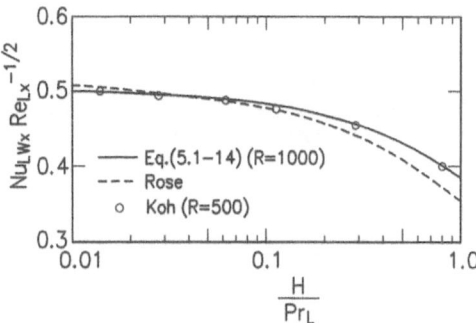

FIGURE 5.1-5. Correlation of $Nu_{Lwx}Re_{Lx}^{-1/2}$ versus H/Pr_L for the range of $Pr_L = 0.003$–0.03 and $R = 1000$.

The similarity solution in the region $H/Pr_L < 0.01$ is correlated by Eq. (5.1-8) for the case of nonmetal vapors. In the case where the pressure of metal vapor is very low, the thermal resistance in the vapor boundary layer becomes dominant owing to the large gaps in temperature and velocity at the vapor-liquid interface, and consequently, the similarity for the condensate film is not exactly valid.

5.2   Forced-Convection Condensation of Superheated Pure Vapors

Taking $T_i = T_s$, $W_1 = 0$, and neglecting the diffusion effect term in Eq. (3.6-1), we obtain

$$0.433 \left(1.367 - \frac{0.432}{\sqrt{2\dot{M}_{FV}}} + \frac{1}{2\dot{M}_{FV}} \right)^{\frac{1}{2}} = \frac{Pr_L \dot{M}_{FV}}{RH}$$

$$+ C_F(Pr_V) \frac{\lambda_V}{\lambda_L} \left(\frac{\nu_L}{\nu_V} \right)^{\frac{1}{2}} \frac{(T_{V\infty} - T_s)}{(T_s - T_w)} \left(1 + 0.26 Pr_V^{0.66} \dot{M}_{FV}^{1.05} \right) , \quad (5.2\text{-}1)$$

where $C_F(Pr_V)$ can be approximated by Eq. (3.4-12b), because Pr_V for a pure vapor varies in the range 0.6–1.

For given values of saturation temperature T_s, degree of superheat $\Delta T_{V\infty} = (T_{V\infty} - T_s)$, and wall temperature T_w, the dimensionless condensate mass flux $\dot{M}_{FV}$ can be calculated from Eq. (5.2-1). The heat flux at the wall q_{wx}, condensation mass flux $\dot{m}_x$, and vapor-layer heat flux at the vapor-liquid interface q_{cx} can then be obtained from the following equations [see Eqs. (3.4-4), (3.4-8), and (3.4-9)]:

$$q_{wx} = 0.433 \left(1.367 - \frac{0.432}{\sqrt{2\dot{M}_{FV}}} + \frac{1}{2\dot{M}_{FV}} \right)^{\frac{1}{2}} (T_s - T_w) \frac{\lambda_L}{x} Re_{Lx}^{\frac{1}{2}} ,$$

$$(5.2\text{-}2)$$

$$\dot{m}_x = \mu_V \frac{Re_{Vx}^{\frac{1}{2}}}{x} \dot{M}_{FV} , \qquad (5.2\text{-}3)$$

$$q_{cx} = q_{wx} - \dot{m}_x \Delta h_V . \qquad (5.2\text{-}4)$$

The effect of the degree of superheat on condensation mass flux and heat flux at the wall is clearly explained through graphical solution of Eq. (5.2-1). For simplicity, the term on the left-hand side of the equation is abbreviated as $Y_{F1}(\dot{M}_{FV})$ and the first and second terms on the right-hand side are $Y_{F2}(\dot{M}_{FV}, T_s - T_w)$ and $Y_{F3}(\dot{M}_{FV}, T_s - T_w, T_{V\infty} - T_s)$, respectively. Figures 5.2-1(a) and (b) show the examples of steam ($T_s = 373.15$ K, $p_s =$

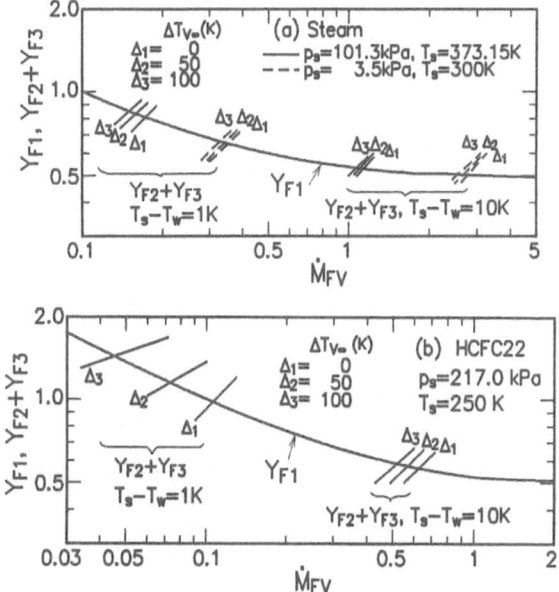

FIGURE 5.2-1. Graphical method for solving Eq. (5.2-1) for forced-convection condensation of a superheated pure vapor. (a) Steam, (b) HCFC22.

101.3 kPa and $T_s = 300$ K, $p_s = 3.5$ kPa) and HCFC22 vapor ($T_s = 250$ K, $p_s = 217$ kPa), respectively, where Y_{F1} and $(Y_{F2} + Y_{F3})$ are represented in the ordinate and $\dot{M}_{FV}$ in the abscissa. The curve moving downwards from the left represents Y_{F1}, while the curves moving downwards from the right represent $Y_{F2} + Y_{F3}$. The intersecting point between the Y_{F1} and $Y_{F2} + Y_{F3}$ curves corresponds to the solution of Eq. (5.2-1), where the ordinate means $Nu_{Lwx}/Re_{Lx}^{1/2}$.

In Figs. 5.2-1(a) and (b) the cases of $T_s - T_w = 1$ K and 10 K and $\Delta T_{V\infty} = 0$, 50 and 100 K are presented. The comparison of superheated vapor cases with saturated vapor cases ($\Delta T_{V\infty} = 0$) at the same pressure and for the same temperature difference $T_s - T_w$ shows that the heat flux q_{wx} increases as the degree of superheat increases while the condensate mass flux $\dot{M}_{FV}$ decreases. This result is explained as follows. The convective heat flux transferred from the vapor phase to the vapor-liquid interface increases as the vapor is superheated. This heat flux is added to the latent heat released at the vapor-liquid interface, and hence the heat flux in the condensate film increases. This increases the temperature gradient in the film, and results in the thinning of the condensate film for a fixed temperature difference $T_s - T_w$. Accordingly, the condensation mass flux decreases. We can see in Figs. 5.2-1(a) and (b) that the degree of increase in q_{wx} and decrease in $\dot{M}_{FV}$ depend on $T_s - T_w$ and bulk pressure as well as $\Delta T_{V\infty}$, which is significant at a small value of $T_s - T_w$ and at high pressure. This is due to the fact that the value of Y_{F2} and Y_{F3} are affected by physical properties. We now quantitatively discuss the effect of superheating upon condensation mass flux and wall heat flux.

We calculate the values of $(\dot{m}_x)_{sup}/(\dot{m}_x)_{sat}$ and $(q_{wx})_{sup}/(q_{wx})_{sat}$ for a fixed value of $(T_s - T_w)$, where $(\dot{m}_x)_{sup}$ and $(q_{wx})_{sup}$ are condensation mass flux and heat flux for a superheated vapor, respectively, while $(\dot{m}_x)_{sat}$ and $(q_{wx})_{sat}$ are those for a saturated vapor. At first, we consider the effect of the difference of physical property $\rho_V \mu_V$ or R between the superheated and saturated vapor upon the above values. For a saturated vapor the following equation is derived from Eqs. (5.1-7), (5.1-8), and (5.1-9):

$$(\dot{M}_{FV})_{sat} = 0.45 \left\{ 1.20 + \left(\frac{Pr_L}{RH} \right)_{sat} \right\}^{\frac{1}{3}} \left(\frac{RH}{Pr_L} \right)_{sat}. \qquad (5.2\text{-}5)$$

Changing $(Pr_L/RH)_{sat}$ in Eq. (5.2-5) to $(Pr_L/RH)_{sup}$ yields

$$(\dot{M}_{FV})_{sat} = 0.45 \left[1.20 + \left\{ \frac{(\rho_V \mu_V)_{sat}}{(\rho_V \mu_V)_{sup}} \right\}^{\frac{1}{2}} \left(\frac{Pr_L}{RH} \right)_{sup} \right]^{\frac{1}{3}}$$
$$\times \left\{ \frac{(\rho_V \mu_V)_{sup}}{(\rho_V \mu_V)_{sat}} \right\}^{\frac{1}{2}} \left(\frac{Pr_L}{RH} \right)_{sup}. \qquad (5.2\text{-}6)$$

We define the following equation by replacing $(Pr_L/RH)_{sat}$ in Eq. (5.2-5) by $(Pr_L/RH)_{sup}$:

$$(\dot{M}_{FV})_{sat/sup} = 0.45 \left\{ 1.20 + \left(\frac{Pr_L}{RH} \right)_{sup} \right\}^{\frac{1}{3}} \left(\frac{RH}{Pr_L} \right)_{sup} \qquad (5.2\text{-}7)$$

and by using this equation, we rewrite Eq. (5.2-6) as

$$(\dot{M}_{FV})_{sat} = \frac{1}{K} \left\{ \frac{(\rho_V \mu_V)_{sup}}{(\rho_V \mu_V)_{sat}} \right\}^{\frac{1}{2}} (\dot{M}_{FV})_{sat/sup}, \qquad (5.2\text{-}8)$$

where

$$K = \left[\frac{1.20 + \left(\dfrac{Pr_L}{RH} \right)_{sup}}{1.20 + \left\{ \dfrac{(\rho_V \mu_V)_{sat}}{(\rho_V \mu_V)_{sup}} \right\}^{\frac{1}{2}} \left(\dfrac{Pr_L}{RH} \right)_{sup}} \right]^{\frac{1}{3}}. \qquad (5.2\text{-}9)$$

$(\dot{M}_{FV})_{sup}/(\dot{M}_{FV})_{sat}$ can be written as

$$\frac{(\dot{M}_{FV})_{sup}}{(\dot{M}_{FV})_{sat}} = \frac{(R\dot{M}_{FL})_{sup}}{(R\dot{M}_{FL})_{sat}} = \left\{ \frac{(\rho_V \mu_V)_{sat}}{(\rho_V \mu_V)_{sup}} \right\}^{\frac{1}{2}} \frac{(\dot{m}_x)_{sup}}{(\dot{m}_x)_{sat}}. \qquad (5.2\text{-}10)$$

Substituting $(\dot{M}_{FV})_{sat}$ of Eq. (5.2-8) into Eq. (5.2-10), and by rearranging it we obtain

$$\frac{(\dot{m}_x)_{sup}}{(\dot{m}_x)_{sat}} = K \frac{(\dot{M}_{FV})_{sup}}{(\dot{M}_{FV})_{sat/sup}} . \qquad (5.2\text{-}11)$$

As for the heat flux q_{wx}, in the same way, by defining

$$(q_{wx})_{sat/sup} = 0.45 \left\{ 1.20 + \left(\frac{Pr_L}{RH} \right)_{sup} \right\}^{\frac{1}{3}} \frac{\lambda_L}{x} Re_{Lx}^{\frac{1}{2}} (T_s - T_w) , \quad (5.2\text{-}12)$$

we can derive

$$\frac{(q_{wx})_{sup}}{(q_{wx})_{sat}} = K \frac{(q_{wx})_{sup}}{(q_{wx})_{sat/sup}} . \qquad (5.2\text{-}13)$$

Numerical values of K are shown in Fig. 5.2-2.

Equation (5.2-1) asymptotically approaches

$$(\dot{M}_{FV})_{sup} = 0.506 \left(\frac{RH}{Pr_L} \right)_{sup} \left\{ 1 + 0.086 \frac{c_{pV}(T_{V\infty} - T_s)}{\Delta h_V} \right\}^{-1}$$

$$\text{for} (\dot{M}_{FV})_{sup} \to \infty , \qquad (5.2\text{-}14)$$

$$(\dot{M}_{FV})_{sup} = 0.856 Pr_V^{\frac{4}{3}} \left(\frac{RH}{Pr_L} \right)_{sup}^2 \left\{ \frac{\Delta h_V}{c_{pV}(T_{V\infty} - T_s)} \right\}^2$$

$$\text{for} (\dot{M}_{FV})_{sup} \to 0 . \qquad (5.2\text{-}15)$$

Dividing Eqs. (5.2-14) and (5.2-15) by the limiting values of Eq. (5.2-7) for $(RH/Pr_L)_{sup} \to \infty$ and 0, respectively, we can derive

$$\frac{(\dot{M}_{FV})_{sup}}{(\dot{M}_{FV})_{sat/sup}} = 1.06 \left\{ 1 + 0.086 \frac{c_{pV}(T_{V\infty} - T_s)}{\Delta h_V} \right\}^{-1}$$

$$\text{for} (\dot{M}_{FV})_{sup} \to \infty , \qquad (5.2\text{-}16)$$

$$\frac{(\dot{M}_{FV})_{sup}}{(\dot{M}_{FV})_{sat/sup}} = 1.90 \left(\frac{Pr_V RH}{Pr_L} \right)_{sup}^{\frac{4}{3}} \left\{ \frac{\Delta h_V}{c_{pV}(T_{V\infty} - T_s)} \right\}^2$$

$$\text{for} (\dot{M}_{FV})_{sup} \to 0 . \qquad (5.2\text{-}17)$$

Based on the shape of the above two equations, we can anticipate that the ratio $(\dot{M}_{FV})_{sup}/(\dot{M}_{FV})_{sat/sup}$ and $(q_{wx})_{sup}/(q_{wx})_{sat/sup}$ can be expressed

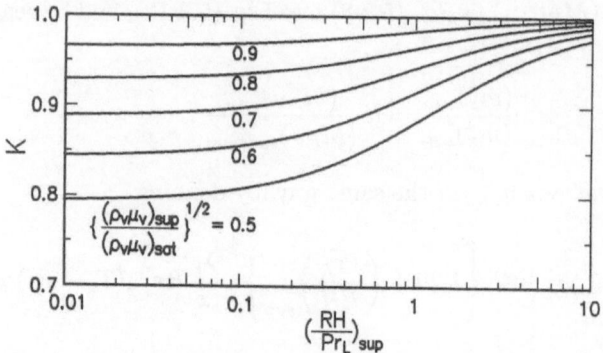

FIGURE 5.2-2. Values of the correction factor K of Eq. (5.2-9).

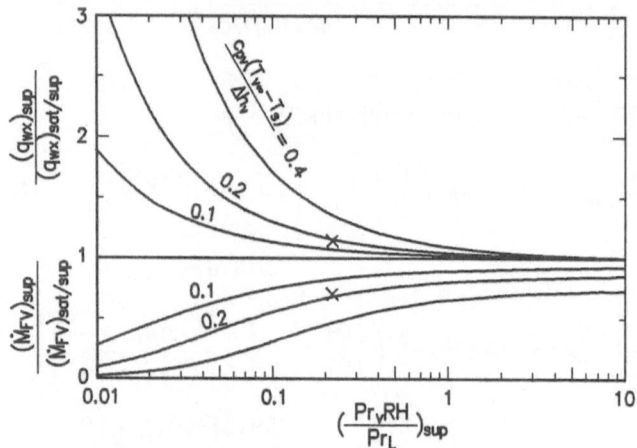

FIGURE 5.2-3. Chart for evaluating the effect of degree of superheat upon forced-convection condensation.

as a function of $(Pr_V RH/Pr_L)_{sup}$ and $c_{pV}(T_{V\infty}-T_s)/\Delta h_V$. The numerical results from Eq. (5.2-1) are shown in Fig. 5.2-3.

In this figure we can see that $(\dot{M}_{FV})_{sup}/(\dot{M}_{FV})_{sat/sup}$ becomes much less than unity while $(q_{wx})_{sup}/(q_{wx})_{sat/sup}$ becomes much more than unity as $(Pr_V RH/Pr_L)_{sup}$ decreases and $c_{pV}(T_{V\infty}-T_s)/\Delta h_V$ increases. The values of $(\dot{m}_x)_{sup}/(\dot{m}_x)_{sat}$ and $(q_{wx})_{sup}/(q_{wx})_{sat}$ are obtained by multiplying $(\dot{M}_{FV})_{sup}/(\dot{M}_{FV})_{sat/sup}$ and $(q_{wx})_{sup}/(q_{wx})_{sat/sup}$ by the value of K in Fig. 5.2-2, respectively. Consequently the values of the formers are closer to unity than those of the latters. When both $(Pr_V RH/Pr_L)_{sup}$ and $c_{pV}(T_{V\infty} - T_s)/\Delta h_V$ are large, $(q_{wx})_{sup}$ is nearly equal to $(q_{wx})_{sat}$ while $(\dot{m}_x)_{sup}$ is appreciably smaller than $(\dot{m}_x)_{sat}$. This characteristic corresponds to a small change in Y_{F1} for large values of $\dot{M}_{FV}$ in Fig. 5.2-1.

Minkowycz and Sparrow[9] solved the two-phase boundary layer equations for a superheated steam ($Pr_V = 0.6$ and 1), graphically presented the result in the relation between q_{sup}/q_{sat} and $(T_s - T_w)$ with $(T_{V\infty} - T_s)$ as a parameter, and found that the effect of superheating appears only for a small value of $(T_s - T_w)$. q_{sup} is larger than q_{sat} by at most ten percent. The point corresponding to $T_s = 100°C$, $T_s - T_w = 1.1°C$, and $T_{V\infty} - T_s = 222.2°C$ in Fig. 2 of Ref. 9 is denoted by the symbol × in Fig. 5.2-3. The value $q_{sup}/q_{sat} \approx 1.15$ at $(Pr_V RH/Pr_L)_{sup} = 0.216$, $c_{pV}(T_{V\infty} - T_s)/\Delta h_V = 0.190$, $(RH/Pr_V)_{sup} = 0.405$, and $(\rho_V\mu_V)_{sat}/(\rho_V\mu_V)_{sup} = 0.96$ agrees well with the result of Minkowycz and Sparrow. It should be noted that $(\dot{m}_x)_{sup}/(\dot{m}_x)_{sat} \approx 0.65$ even in this case.

5.3 Free-Convection Condensation of Saturated Pure Vapors

Substituting $T_{V\infty} = T_i = T_s$ into Eq. (4.6-1), we obtain

$$\dot{M}_{GL} = \left(\frac{H}{Pr_L}\right)^{\frac{3}{4}}. \tag{5.3-1}$$

From Eqs. (4.4-1), (4.4-4), and (5.3-1) the following equation, i.e., Nusselt's[10] equation is derived as

$$Nu_{Lwx} = \frac{1}{\sqrt{2}}\left(\frac{Ga_x Pr_L}{H}\right)^{\frac{1}{4}}. \tag{5.3-2}$$

The Nusselt number averaged over $0 \le x \le \ell$ is expressed as

$$\overline{Nu}_{Lw\ell} = \frac{4}{3\sqrt{2}}\left(\frac{Ga_\ell Pr_L}{H}\right)^{\frac{1}{4}}. \tag{5.3-3}$$

The main assumptions of Nusselt's theory are as follows: (i) the motion of vapor phase does not affect the motion of condensate, which corresponds to $R \to \infty$; (ii) the inertia term in the momentum equation for the condensate film can be neglected; and (iii) the convection term in the energy equation for the condensate film can be neglected. When these assumptions are introduced in the similarity transformation in Section 4.1, the following ordinary differential equations are derived :

$$F_{GL}''' + 1 = 0, \tag{5.3-4}$$
$$\Theta_{GL}'' = 0 \tag{5.3-5}$$

together with the boundary conditions

at $\eta_{GL} = 0$:

$$F_{GLw} = 0, \tag{5.3-6a}$$
$$F'_{GLw} = 0, \tag{5.3-6b}$$
$$\Theta_{GLw} = 1; \tag{5.3-6c}$$

at $\eta_{GL} = \eta_{GLi}$:

$$F''_{GLi} = 0, \tag{5.3-7a}$$
$$\Theta_{GLi} = 0. \tag{5.3-7b}$$

The analytical solution of these equations are

$$F_{GL} = \frac{\eta_{GLi}}{2}\eta_{GL}^2 - \frac{1}{6}\eta_{GL}^3, \tag{5.3-8}$$

$$\Theta_{GL} = 1 - \frac{1}{\eta_{GLi}}\eta_{GL}. \tag{5.3-9}$$

Note that Eq. (5.3-8) is equal to Eq. (4.2-1) when the third through last terms on the right-hand side of Eq. (4.2-1) are neglected. Substituting F_{GLi} and Θ'_{GLi}, which are obtained from Eqs. (5.3-8) and (5.3-9), into the equation

$$\frac{3F_{GLi}}{-\Theta'_{GLi}} = \frac{H}{Pr_L}, \tag{5.3-10}$$

which is derived from Eq. (4.1-24), we obtain

$$-\frac{1}{\Theta'_{GLi}} = \eta_{GLi} = \left(\frac{H}{Pr_L}\right)^{\frac{1}{4}}. \tag{5.3-11}$$

Substituting this equation into Eq. (4.4-1), we can obtain Nusselt's equation (5.3-2).

Sparrow and Gregg[11] numerically obtained the similarity solution of the equation of condensate film, where the shear stress at the vapor-liquid interface is neglected. Chen[12] transformed the basic equations to modified integral equations, and solved them by means of the perturbation method without neglecting the shear stress at the vapor-liquid interface. Later, Maekawa and Rose[13] reexamined this solution and proposed an equation, where thw Nusselt number is accurately approximated as a function of H, H/Pr_L, and Pr_L. All of the above results reveal that Nusselt's equation is valid in error by less than two percent in the normal practical range of $H/Pr_L \lesssim 0.1$.

Figure 5.3-1 shows the similarity solution by Koh et al.[14] for $Pr_L = 0.003$–0.03 in the relation of $Nu_{Lwx}/(Nu_{Lwx})_{Nu}$ versus H/Pr_L, where $(Nu_{Lwx})_{Nu}$ denotes Eq. (5.3-2). The tendency that Nu_{Lwx} decreases as H/Pr_L increases is due to the effect of the inertia term. Note that the value of H/Pr_L for liquid metals is at most 0.5 in practice.

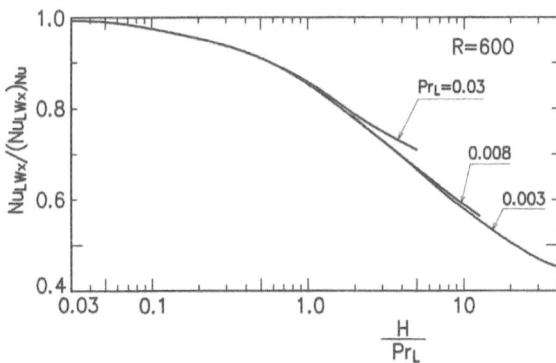

FIGURE 5.3-1. Correlation of $Nu_{Lwx}/(Nu_{Lwx})_{Nu}$ versus H/Pr_L for the ranges of H/Pr_L=0.03–40 and Pr_L=0.003–0.03 in the case of free-convection condensation.

5.4   Free-Convection Condensation of Superheated Pure Vapors

When $T_i = T_s$ and $W_1 = 0$, Eq. (4.6-1) reduces to

$$\dot{M}_{GL}^{-\frac{1}{3}} = \frac{Pr_L \dot{M}_{GL}}{H} + \sqrt{2} C_G (Pr_V) \frac{\lambda_V}{\lambda_L} \left(\frac{\nu_L}{\nu_V} \right)^{\frac{1}{2}} \frac{T_{V\infty} - T_s}{T_s - T_w}$$

$$\times \left\{ \frac{T_{V\infty} - T_s}{T_{V\infty}} Pr_V \right\}^{\frac{1}{4}} \left[1 + \left\{ \left(\frac{T_{V\infty}}{T_{V\infty} - T_s} \right)^{\frac{1}{4}} R\dot{M}_{GL} \right\}^{1.2} \right], \quad (5.4\text{-}1)$$

where $C_G(Pr_V)$ can be approximated by Eq. (4.4-6b) for the range $Pr_V = 0.6$–1.

Dimensionless condensation mass flux $\dot{M}_{GL}$ for specified values of T_s or p_s, $(T_{V\infty}-T_s)$, and T_w is solved from Eq. (5.4-1), and then q_{wx}, $\dot{m}_x$, and q_{cx} are calculated from the following equations :

$$q_{wx} = \frac{\lambda_L (T_s - T_w)}{x} \dot{M}_{GL}^{-\frac{1}{3}} \left(\frac{Ga_x}{4} \right)^{\frac{1}{4}}, \quad (5.4\text{-}2)$$

$$\dot{m}_x = \frac{\mu_L}{x} \dot{M}_{GL} \left(\frac{Ga_x}{4} \right)^{\frac{1}{4}}, \quad (5.4\text{-}3)$$

$$q_{cx} = q_{wx} - \dot{m}_x \Delta h_V. \quad (5.4\text{-}4)$$

To clarify the mechanism of the effect of degree of superheat, a method of the graphical solution of Eq. (5.4-1) is explained. Similarly, as in Section 5.2, the term on the left-hand side of the equation is abbreviated as $Y_{G1}(\dot{M}_{GL})$ and the first and second terms on the right-hand side are $Y_{G2}(\dot{M}_{GL}, T_s-T_w)$ and $Y_{G3}(\dot{M}_{GL}, T_s-T_w, T_{V\infty}-T_s, T_{V\infty})$, respectively. As

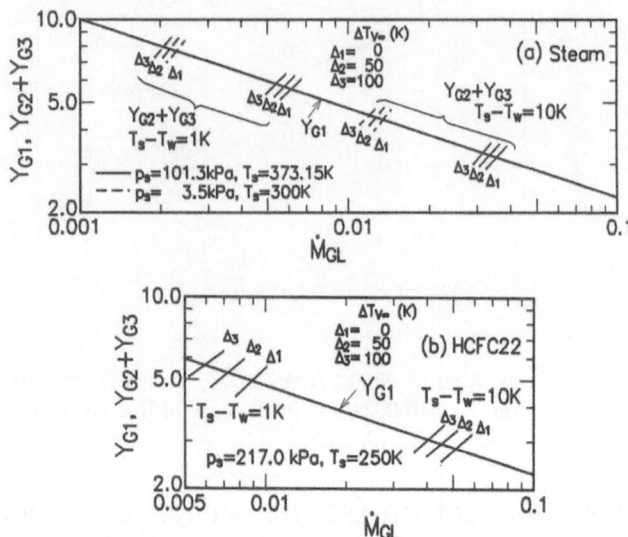

FIGURE 5.4-1.   Graphical method for solving Eq. (5.4-1) for free-convection condensation of a superheated pure vapor. (a) Steam, (b) HCFC22.

for the same vapors with the same conditions of pressure and temperature as in the case of forced-convection condensation, the values of Y_{G1} and $Y_{G2} + Y_{G3}$ are plotted against $\dot{M}_{GL}$ in Figs. 5.4-1(a) for steam and (b) for HCFC22, where the straight line moving downwards from left represents Y_{G1} and the curves moving downwards from the right represent $Y_{G2} + Y_{G3}$. The intersecting point between Y_{G1} and $Y_{G2} + Y_{G3}$ curves corresponds to the solution of Eq. (5.4-1), where the ordinate means $Nu_{Lwx}/(Ga_x/4)^{1/4}$.

In comparison of Figs. 5.4-1(a) and (b) with Figs. 5.2-1(a) and (b), we can find that the decrease of $\dot{m}_x$ and increase of q_{wx} due to superheating and the effects of temperature and pressure upon them are similar to the case of forced-convection condensation. We now discuss these characteristics quantitatively.

Since the second term on the right-hand side of Eq. (5.4-1) is smaller than the first term, the equation can be transformed as

$$
\frac{(\dot{M}_{GL})_{sup}}{(\dot{M}_{GL})_{sat}} = \left[1 + \frac{0.566 Pr_V^{0.1}}{Pr_V^{\frac{3}{4}}} \frac{c_{pV}(T_{V\infty} - T_s)}{\Delta h_V} \frac{\left(\frac{T_{V\infty} - T_s}{T_{V\infty}} \right)^{\frac{1}{4}}}{R \left(\frac{H}{Pr_L} \right)^{\frac{3}{4}}} \right.
$$

$$
\left. \times \frac{(\dot{M}_{GL})_{sat}}{(\dot{M}_{GL})_{sup}} \left\{ 1 + \left(\frac{R \left(\frac{H}{Pr_L} \right)^{\frac{3}{4}}}{\left(\frac{T_{V\infty} - T_s}{T_{V\infty}} \right)^{\frac{1}{4}}} \frac{(\dot{M}_{GL})_{sup}}{(\dot{M}_{GL})_{sat}} \right)^{1.2} \right\} \right]^{-\frac{3}{4}} ,
$$

$$(5.4\text{-}5)$$

where the physical properties of superheated vapor are taken. It is anticipated from Eq. (5.4-5) that $(\dot{M}_{GL})_{sup}/(\dot{M}_{GL})_{sat}$ will be approximately expressed as a function of $c_{pV}(T_{V\infty}-T_s)/\Delta h_V$ and the following parameter:

$$
\xi = \left(\frac{1}{Pr_V^{0.65} R} \right) \left(\frac{Pr_L}{H} \right)^{\frac{3}{4}} \left(\frac{T_{V\infty} - T_s}{T_{V\infty}} \right)^{\frac{1}{4}} . \tag{5.4-6}
$$

Figures 5.4-2(a) and (b) show the solution of Eq. (5.4-1) in the coordinate of $(\dot{M}_{GL})_{sup}/(\dot{M}_{GL})_{sat}$ versus $c_{pV}(T_{V\infty}-T_s)/\Delta h_V$ with ξ as a parameter and in the coordinate of $(\dot{M}_{GL})_{sup}/(\dot{M}_{GL})_{sat}$ versus ξ with $c_{pV}(T_{V\infty}-T_s)/\Delta h_V$ as a parameter, respectively.

As for the heat flux at the cooling surface, the following equation can be derived:

$$
\frac{(q_{wx})_{sup}}{(q_{wx})_{sat}} = \left\{ \frac{(\dot{m}_x)_{sup}}{(\dot{m}_x)_{sat}} \right\}^{-\frac{1}{3}} . \tag{5.4-7}
$$

Sparrow and Eckert[15] proposed the following equation by an approximate analysis, in which the heat transfer due to convection in the vapor layer is neglected:

$$
\frac{(q_{wx})_{sup}}{(q_{wx})_{sat}} = \left\{ 1 + \frac{c_{pV}(T_{V\infty} - T_s)}{\Delta h_V} \right\}^{\frac{1}{4}} . \tag{5.4-8}
$$

An equation of $(\dot{M}_{GL})_{sup}/(\dot{M}_{GL})_{sat}$ can be derived by substituting Eq. (5.4-8) into Eq. (5.4-7). The equation thus obtained agrees with the curve for $\xi = 0.1$ in Fig. 5.4-2(a). This means that the result of Sparrow and Eckert[15] corresponds to the case for a small value of ξ, i.e., for a large value of (T_s-T_w) and a small value of $(T_{V\infty}-T_s)$.

Minkowycz and Sparrow[16] showed the effect of degree of superheat on free-convection condensation of steam, which were numerically solved by

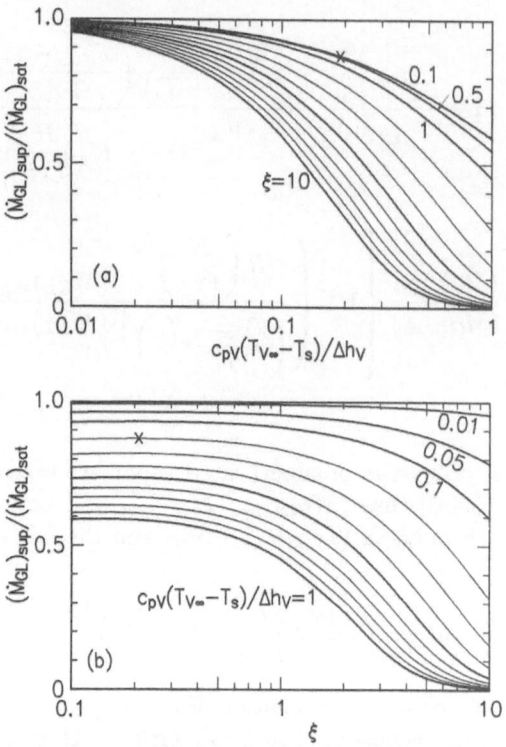

FIGURE 5.4-2.   Chart for evaluating the effect of degree of super-heat upon free-convection condensation.   (a) $(\dot{M}_{GL})_{sup}/(\dot{M}_{GL})_{sat}$ versus $c_{pV}(T_{V\infty} - T_s)/\Delta h_V$. (b) $(\dot{M}_{GL})_{sup}/(\dot{M}_{GL})_{sat}$ versus ξ.

considering the variation of physical properties. The result represents that q_{sup}/q_{sat} for a fixed $(T_{V\infty}-T_w)$ value is almost independent of T_s and (T_s-T_w). They also showed that the value $q_{sup}/q_{sat} = 1.047$ for $T_s = 373.15$ K and $T_{V\infty}-T_w = 222$ K agrees with $q_{sup}/q_{sat} = 1.045$ from Eq. (5.4-8) of Sparrow and Eckert.[15] The symbol $\times$ in Figs. 5.4-2(a) and (b) corresponds to the value at $T_s = 373.15$ K, $T_{V\infty}-T_w = 222$ K, and $T_s-T_w = 5$ K, where $c_{pV}(T_{V\infty}-T_s)/\Delta h_V = 0.190$ and $\xi = 0.212$. From Fig. (5.4-2) or Eq. (5.4-5) we obtain $(\dot{m}_x)_{sup}/(\dot{m}_x)_{sat} = 0.87$, further from Eq. (5.4-7) we obtain $q_{sup}/q_{sat} = 1.048$, which agrees with the above two values. These values show that the effect of degree of superheat appears more remarkable in $\dot{m}_x$ than q_{wx} as in forced-convection condensation.

Minkowycz and Sparrow[16] also reported that the effect of superheating upon the heat flux at the wall is more significant in free-convection condensation than in forced-convection condensation. However, this is not generally valid as seen by comparing Figs. 5.2-1(a) and (b) and 5.4-1(a) and (b) for a water vapor ($p_s = 101.3$ kPa) and a HCFC22 vapor ($p_s =$

217 kPa). As for $T_s - T_w = 1$ K, the effect of degree of superheat is more significant in forced-convection condensation while as for $T_s - T_w = 10$ K it is more significant in free-convection condensation. Generally speaking, the effect of superheating is significant in the case of thin condensate film.

Traditionally, condensers in chemical plants are designed under the assumption that superheated vapor does not condense, and accordingly a desuperheating zone is usually set at the entrance region of a condenser. This idea is approximately correct in the case of an extremely small value of $(\dot{m}_x)_{sup}/(\dot{m}_x)_{sat}$, as seen in Figs. 5.2-3 and 5.4-2. Also, it should be noted that the heat flux q_{wx} in this case is significantly larger than that in single phase heat transfer.

5.5 Combined Forced- and Free-Convection Condensation of Saturated Pure Vapors

On a vertical surface, forced-convection condensation takes place in the case of a large value of $U_{V\infty}$ and free-convection condensation in the case of a small value of $U_{V\infty}$. For a medium value of $U_{V\infty}$ the characteristic of forced-convection condensation appears near the top of the surface, while that of free-convection condensation appears near the bottom of the surface. This case is named combined forced- and free-convection condensation, for which no similarity solution exists. Jacobs[17] obtained a numerical solution by using the Kármán-Pohlhausen integral method, and demonstrated the result in graphs. Shekriladze and Gomelauri[18] derived an equation of Nu as a function of Re_L, H, and Fr by applying their unique approximate method as explained in the next section. In this section we explain the result of Fujii and Uehara,[2] which was obtained by correcting Jacob's method.

Fujii and Uehara[2] derived simultaneous ordinary differential equations with respect to the condensate film thickness δ and the vapor velocity U_i at the vapor-liquid interface by neglecting the inertia and convection terms in the condensate film and by assuming a velocity profile in the vapor boundary layer, and then solving them numerically. The result for heat transfer is shown by solid lines in the coordinates of $Nu_{Lwx}/Re_{Lx}^{1/2}$ versus $Pr_L/Fr_x H$ in Fig. 5.5-1. The solid lines in the figure can be approximated by the following equation within the error of 2.5 percent :

$$Nu_{Lwx} = \left\{ \left(\chi Re_{Lx}^{\frac{1}{2}} \right)^4 + \frac{Ga_x Pr_L}{4H} \right\}^{\frac{1}{4}} \tag{5.5-1a}$$

$$= \chi \left\{ 1 + \frac{Pr_L}{4\chi^4 Fr_x H} \right\}^{\frac{1}{4}} Re_{Lx}^{\frac{1}{2}}, \tag{5.5-1b}$$

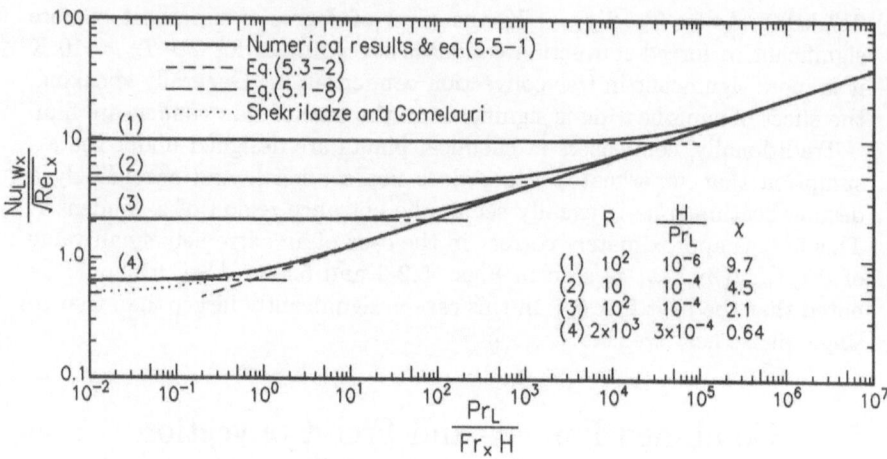

FIGURE 5.5-1. Correlation of $Nu_{Lwx}/Re_{Lx}^{1/2}$ versus $Pr_L/Fr_x H$ in the case of combined forced- and free-convection condensation of a pure vapor.

where

$$Fr_x = \frac{U_{V\infty}^2}{gx}. \tag{5.5-2}$$

Equation (5.5-1) is the (1/4)th power of the sum of the 4th power of Eq. (5.1-8) and the 4th power of Eq. (5.3-2): this is named the 4th power combination rule. In Fig. 5.5-1 the dot-dash lines represent Eq. (5.1-8) and a broken line represents Eq. (5.3-2). We may roughly consider the intersecting points of the boundary between the forced-convection region and the free-convection region. The points expressed by x_c are

$$x_c = \frac{4\chi^4 U_{V\infty}^2 H}{g Pr_L}. \tag{5.5-3}$$

The average Nusselt number in the range $0 \le x \le \ell$ is approximately expressed as

$$\overline{Nu}_{Lw\ell} = \left\{ 0.655 \left(1.20 + \frac{Pr_L}{RH} \right)^{\frac{4}{3}} Re_{L\ell}^2 + 0.79 \frac{Ga_\ell Pr_L}{H} \right\}^{\frac{1}{4}} \tag{5.5-4a}$$

$$= 2\chi \left\{ 1 + \left(\frac{\sqrt{2}}{3\chi} \right)^4 \frac{Pr_L}{Fr_\ell H} \right\}^{\frac{1}{4}} Re_{L\ell}^{\frac{1}{2}}, \tag{5.5-4b}$$

where the subscript ℓ denotes the value at $x = \ell$. The boundary between the forced-convection region and the free-convection region for the average

Nusselt number is expressed as

$$(x_c)_m \approx \frac{20\chi^4 U_{V\infty}^2 H}{g \Pr_L} .$$ (5.5-5)

In the comparison between Eqs. (5.5-3) and (5.5-5) we recognize that the appearance of the characteristic of the free-convection condensation in $\overline{Nu_{Lw\ell}}$ is retarded by five times in the x position.

The dominant feature of condensation on a vertical surface of length ℓ can be postulated by the criterion

$$\left.\begin{array}{lll}
\text{free-convection condensation for} & (x_c)_m/\ell \lesssim 0.1 \\
\text{combined-convection condensation for} & 0.1 \lesssim (x_c)_m/\ell \lesssim 0.9 \\
\text{forced-convection condensation for} & 0.9 \lesssim (x_c)_m/\ell
\end{array}\right\} . \quad (5.5\text{-}6)$$

5.6 Discussion on the Shekriladze and Gomelauri's Solution for Forced-Convection Condensation of Saturated Pure Vapors

Shekriladze and Gomelauri[18] considered that the shear stress at the vapor-liquid interface is due to the momentum change between the vapor in the free stream and condensate surface, i.e.,

$$\tau_w = \tau_i = \dot{m}_x (U_{V\infty} - U_{Vi}) .$$ (5.6-1)

Using this equation they derived the following equation for the forced-convection condensation:

$$Nu_{Lwx} = \frac{1}{2}\left(1 + \frac{H}{\Pr_L}\right)^{-\frac{1}{2}} Re_{Lx}^{\frac{1}{2}}$$ (5.6-2)

for the case of $U_{V\infty} \gg U_{Vi}$.

Equation (5.6-1) can be derived by assuming $\partial V_V/\partial y = 0$ in Eq. (2-4) (continuity equation of vapor phase), i.e., $V_V = V_V(x)$, by integrating Eq. (2.5) (momentum equation of vapor phase for $g = 0$) with respect to y, and by approximating $-\rho_V V_{Vi} = \dot{m}_x$. Since Eq. (5.6-2) is widely used for condensation studies, the relation between this equation and Eq. (5.1-8) is clarified below.

Equation (3.7-2) can be transformed by using Eqs. (3.1-22), (3.7-8), (3.8-22), (3.4-4), and (5.1-8) as

$$\tau_w \approx \tau_i = \mu_L \left(\frac{\partial U_L}{\partial y}\right)_i = 4\chi^2 \dot{m}_x U_{V\infty} .$$ (5.6-3)

Using this equation instead of Eq. (5.6-1), we explain the method of solution by Shekriladze and Gomelauri.

Solving Eq. (2-2), in which the inertia term is neglected and g is taken to be zero, and using Eq. (5.6-3) we obtain

$$\frac{\partial U_L}{\partial y} = \frac{4\chi^2}{\mu_L}\dot{m}_x U_{V\infty} \, . \tag{5.6-4}$$

The integration of Eq. (5.6-4) under the condition of $U_L = 0$ at $y = 0$ yields

$$U_L = \frac{4\chi^2 \dot{m}_x U_{V\infty}}{\mu_L} y \, . \tag{5.6-5}$$

Substituting Eq. (5.6-5) into Eq. (2-35), we obtain

$$\dot{m}_x = \frac{U_{V\infty}}{\nu_L} \frac{d}{dx}\left(2\chi^2 \dot{m}_x \delta^2\right) \, . \tag{5.6-6}$$

Substituting the following equation, which corresponds to linear temperature distribution,

$$\dot{m}_x = \frac{q_{wx}}{\Delta h_V} = \frac{\lambda_L (T_s - T_w)}{\delta \Delta h_V} \, , \tag{5.6-7}$$

into both sides of Eq. (5.6-6) and rearranging yields

$$\frac{d\delta^2}{dx} = \frac{\nu_L}{\chi^2 U_{V\infty}} \, . \tag{5.6-8}$$

δ is obtained by integrating Eq. (5.6-8) under the condition of $\delta = 0$ at $x = 0$, and Nu_{Lwx} is then expressed as

$$\mathrm{Nu}_{Lwx} = \frac{q_{wx} x}{(T_s - T_w)\lambda_L} = \frac{x}{\delta} = \chi \mathrm{Re}_{Lx}^{\frac{1}{2}} \, . \tag{5.6-9}$$

This equation coincides with Eq. (5.1-8). In conclusion, the method of Shekliladze and Gomelauri is very simple and useful if the value of τ_w is given accurately. Their original solution [Eq. (5.6-2)] corresponds to the case of $4\chi^2 = 1$ or the shear stress at $RH/\mathrm{Pr}_L \approx 5$.

5.7   Condensation of Saturated Pure Vapors in the Case of Uniform Heat Flux

The condition of cooling surfaces in actual condensers is neither uniform temperature nor uniform heat flux. However, we can more or less predict the actual heat transfer characteristics, if we have the knowledge of these two extreme cases. This is the reason why we discuss the condensation for the case of uniform heat flux in this section, although we cannot find its similarity solution.

5.7.1 FORCED-CONVECTION CONDENSATION

The characteristics of forced-convection condensation of a saturated vapor in the case of uniform surface heat flux q_0 can be derived by means of the method of Shekriladze and Gomelauri.[18] Under the assumptions that the shear stress at the vapor-liquid interface is expressd by Eq. (5.6-3) and $\dot{m}$ is independent of x, the following equation is derived from Eq. (5.6-6)

$$\frac{d\delta^2}{dx} = \frac{\nu_L}{2\chi^2 U_{V\infty}}. \tag{5.7-1}$$

This equation is integrated under the condition $\delta = 0$ at $x = 0$ as

$$\delta = \left\{\frac{\nu_L x}{2\chi^2 U_{V\infty}}\right\}^{\frac{1}{2}}. \tag{5.7-2}$$

This means that the variation of δ in the x direction is similar to the case of uniform surface temperature.

The variation of surface temperature and the local Nusselt number are expressed as

$$T_s - T_{wx} = \frac{q_0\delta}{\lambda_L} = \frac{q_0 x}{\sqrt{2}\chi\lambda_L}Re_{Lx}^{-\frac{1}{2}} \propto x^{\frac{1}{2}}, \tag{5.7-3}$$

$$Nu_{Lwx} = \frac{q_0 x}{(T_s - T_{wx})\lambda_L} = \sqrt{2}\chi Re_{Lx}^{\frac{1}{2}}. \tag{5.7-4}$$

The average Nusselt number based on the area average surface temperature is expressed as

$$\overline{Nu}_{Lw\ell} = \frac{3}{\sqrt{2}}\chi Re_{Lx}^{\frac{1}{2}}. \tag{5.7-5}$$

The average heat transfer coefficient from this equation coincides with the local heat transfer coefficient at $x = \ell/2$.

5.7.2 FREE-CONVECTION CONDENSATION[19]

We can solve the problem of free-convection condensation of a saturated vapor for the case of uniform surface heat flux q_0 by introducing Nusselt's assumptions.

Basic equations are written as

$$\nu_L\frac{\partial^2 U_L}{\partial y^2} + g = 0, \tag{5.7-6}$$

$$\lambda_L\frac{\partial T_L}{\partial y} = q_0, \tag{5.7-7}$$

$$\rho_L\Delta h_V\frac{d}{dx}\int_0^\delta U_L dy = q_0 \tag{5.7-8}$$

together with the boundary conditions

at $y = 0$: $U_L = 0$;   (5.7-9)

at $y = \delta$: $\left\{ \dfrac{\partial U_L}{\partial y} \right\}_\delta = 0$,   (5.7-10)

$$T_L = T_s ;$$ (5.7-11)

where Eq. (5.7-8) is derived by substituting Eqs. (5.7-7) and (2-35) into Eq. (2-25).

The solution of Eq. (5.7-6) is

$$U_L = \frac{g}{\nu_L} \left(\delta y - \frac{1}{2} y^2 \right) .$$ (5.7-12)

Solving Eq. (5.7-8), into which Eq. (5.7-12) is substituted, subject to the conditions of $\delta = 0$ at $x = 0$, we obtain

$$\delta = \left(\frac{3 \nu_L q_0 x}{g \rho_L \Delta h_V} \right)^{\frac{1}{3}} .$$ (5.7-13)

On the other hand, the solution of Eq. (5.7-7) is

$$T_s - T_L = \frac{q_0}{\lambda_L} (\delta - y) .$$ (5.7-14)

Substituting δ of Eq. (5.7-13) into Eq. (5.7-14), we obtain the cooling surface temperature T_w as

$$T_s - T_{wx} = \left(\frac{3 \nu_L q_0^4 x}{g \rho_L \lambda_L^3 \Delta h_V} \right)^{\frac{1}{3}} .$$ (5.7-15)

Both δ and $(T_s - T_{wx})$ increase in proportion to $x^{1/3}$ in the case of uniform surface heat flux. Note that δ and q_{wx} are proportional to $x^{1/4}$ and $x^{-1/4}$, respectively, in the case of uniform surface temperature.

We define the local heat transfer coefficient α_{wx} as

$$\alpha_{wx} = \frac{q_0}{T_s - T_{wx}} = \left(\frac{3 \nu_L q_0 x}{g \rho_L \lambda_L^3 \Delta h_V} \right)^{-\frac{1}{3}} .$$ (5.7-16)

This equation is nondimensionalized as

$$Nu_x^* = \left(\frac{3}{4} Re_x^* \right)^{-\frac{1}{3}} ,$$ (5.7-17)

where

$$Nu_x^* = \frac{q_0 \left(\dfrac{\nu_L^2}{g}\right)^{\frac{1}{3}}}{(T_s - T_{wx})\lambda_L}, \tag{5.7-18}$$

$$Re_x^* = \frac{4q_0\, x}{\mu_L \Delta h_V}. \tag{5.7-19}$$

When we define the average heat transfer coefficient $\overline{\alpha_w}$ over the cooling surface length ℓ based on the area average temperature $\overline{T_w}$ we obtain

$$\overline{Nu^*} = \frac{q_0 \left(\dfrac{\nu_L^2}{g}\right)^{\frac{1}{3}}}{(T_s - \overline{T_w})\lambda_L} = \frac{4}{3}\left(\frac{3}{4}Re_\ell^*\right)^{-\frac{1}{3}}. \tag{5.7-20}$$

Since $(\nu_L^2/g)^{1/3}$ in Eqs. (5.7-18) and (5.7-20) has the dimension of length, Nu^* and $\overline{Nu^*}$ are named modified Nusselt numbers. Since $q_0 x/\Delta h_V$ in Eq. (5.7-19) represents the mass flow rate Γ_x per unit width of cooling surface at x, Re_x^* can be rewritten as

$$Re_x^* = \frac{4\Gamma_x}{\mu_L} = \frac{\overline{U}_{Lx}(4\delta)}{\nu_L}, \tag{5.7-21}$$

where $\overline{U}_{Lx}$ denotes the mean velocity of condensate at x and 4δ is the mean hydraulic diameter of the cross section of condensate film at x. Re_x^* is therefore named the film Reynolds number.

We can derive Eq. (5.7-20) from Eq. (5.3-3) by replacing $\overline{q_w}$ in Eq. (5.3-3) by q_0. Consequently, if we take the area average values of $\overline{T_w}$ and $\overline{q_w}$, the average heat transfer coefficient can be expressed by Eq. (5.7-20) irrespective of the cooling surface condition. This is one of the reasons why Eq. (5.7-20) is used for correlating experimental results on free-convection condensation. However, it should be kept in mind that Eq. (5.7-17) is related to the local Nusselt number for calculating the cooling surface temperature distribution for a given value of q_0.

The position x_r at which the local Nusselt number has the same value as the average Nusselt number calculated from Eq. (5.7-20) is $x_r = (3/4)^3 \ell = 0.42\ell$. If we use the temperature difference $(T_s - T_{wx})$ at $x_r = \ell/2$ in the calculation of the average heat transfer coefficient, we obtain a 5.5 percent smaller value than that from Eq. (5.7-20). The area average value of local heat transfer coefficient, though not used in practice, predicts a 12.5 percent larger value than that from Eq. (5.7-20). The above discrepancy should be noted in the comparison between theory and experiment.

REFERENCES

1. Rose, J. W. , Fundamentals of Condensation Heat Transfer: Laminar Film Condensation, *JSME International Journal*, Ser. II, 31, 3, 357–375 (1988).

2. Fujii, T. and H. Uehara, Laminar Filmwise Condensation on a Vertical Surface, *Int. J. Heat Mass Transfer*, 15, 2, 217–233 (1972).

3. Cess, R. D., Laminar-Film Condensation on a Flat Plate in the Absence of a Body Force, *Z. Angew. Math. Phys.*, XI, 426–433 (1960).

4. Sparrow, E. M., W. J. Minkowycz, and M. Saddy, Forced Convection Condensation in the Presence of Noncondensables and Interfacial Resistance, *Int. J. Heat Mass Transfer*, 10, 1829–1845 (1967).

5. Koh, J. C. Y. , Film Condensation in a Forced-Convection Boundary-Layer Flow, *Int. J. Heat Mass Transfer*, 5, 941–954 (1962).

6. Fujii, T. and J. B. Lee, Similarity Solution of Forced Convection Condensation of a Saturated Vapor—For the case of Low $\rho\mu$ Ratio and High Prandtl Number (in preparation).

7. Koh, J. C. Y., Laminar Film Condensation of Condensible Gases and Gaseous Mixtures on a Flat Plate, *Proc. 4th U.S. National Congress of Applied Mechanics*, 2, 1327–1336 (1962).

8. Rose, J. W., A New Interpolation Formula for Forced-Convection Condensation on a Horizontal Surface, *Trans. ASME, J. Heat Transfer*, 111, 818–819 (1989).

9. Minkowycz, W. J. and E. M. Sparrow, The Effect of Superheating on Condensation Heat Transfer in a Forced Convection Boundary Layer Flow, *Int. J. Heat Mass Transfer*, 12, 147–154 (1969).

10. Nusselt, W., Die Oberflachenkondensation des Wasserdampfes, *Z. Ver. Deut. Ing.*, 60, 541–580 (1916).

11. Sparrow, E. M. and J. L. Gregg, A Boundary-Layer Treatment of Laminar-Film Condensation, *Trans. ASME, J. Heat Transfer*, 81, 13–18 (1959).

12. Chen, M. M., An Analytical Study of Laminar Film Condensation: Part 1-Flat Plates, *Trans. ASME, J. Heat Transfer*, 83, 48–54 (1961).

13. Maekawa, T. and J. W. Rose (in preparation, see 1).

14. Koh, J. C. Y., E. M. Sparrow, and J. P. Hartnett, The Two Phase Boundary Layer in Laminar Film Condensation, *Int. J. Heat Mass Transfer*, 2, 69–82 (1961).

15. Sparrow, E. M. and E. R. G. Eckert, Effects of Superheated Vapor and Noncondensable Gases on Laminar Film Condensation, *AIChE J.*, 7, 3, 473–477 (1961).

16. Minkowycz, W. J. and E. M. Sparrow, Condensation Heat Transfer in the Presence of Noncondensables, Interfacial Resistance, Superheating, Variable Properties, and Diffusion, *Int. J. Heat Mass Transfer*, 6, 1125–1144 (1966).

17. Jacobs, H. R., An Integral Treatment of Combined Body Force and Forced Convection in Laminar Film Condensation, *Int. J. Heat Mass Transfer*, 9, 637–648 (1966).

18. Shekriladze, I. G. and V. I. Gomelauri, Theoretical Study of Laminar Film Condensation of Flowing Vapour, *Int. J. Heat Mass Transfer*, 9, 581–591 (1966).

19. Fujii, T., H. Uehara, and K. Oda, Filmwise Condensation on a Surface with Uniform Heat Flux and Body Force Convection, *Heat Transfer Japanese Research*, 1, 4 , 76–83 (1972).

6

Condensation of Binary Vapors

Chapters 3 and 4 presented a method of similarity analysis of a binary vapor mixture along with its solution and an algebraic method for obtaining condensation mass flux and heat flux. In this chapter the latter method is applied to some special cases and the results are considered.

6.1 Forced-Convection Condensation of Mixtures of Vapor and Noncondensable Gas

Sparrow et al.[1] obtained a similarity solution for an air-steam mixture, where the inertia term in Eq. (2-2), the convection term in Eq. (2-3), and Eq. (2-6) for the conservation of energy in a vapor phase were neglected and the velocity at the vapor-liquid interface was assumed to be zero. Their results were shown in the figures of η_{FLi}^{-1} versus RH/Pr_L and $W_{1V\infty}/W_{1Vi}$ versus RH/Pr_L, and in a table containing F_{FLw}, F''_{FLw}, RH/Pr_L, η_{FLi}^{-1}, and $W_{1V\infty}/W_{1Vi}$. By using these figures they calculated the heat flux q_{wx} and graphically showed the ratio of q_{wx} to the heat flux $(q_{wx})^0$ for a pure steam against $(T_{V\infty} - T_w)$ with $W_{1V\infty}$ and $T_{V\infty}$ as parameters. Further, they clarified that the heat transfer is scarcely affected by the assumed temperature jump at the vapor-liquid interface.

By means of a similar numerical analysis, Minkowycz and Sparrow[2] investigated the effect of superheating of bulk vapor upon the heat transfer in forced-convection condensation of an air-steam mixture. Their results are summarized in tables and figures. By using these data they graphically showed the ratio of the heat flux $q_{wx,sup}$ for a superheated mixture to the heat flux $(q_{wx,sup})^0$ for a saturated mixture against $(T_{s\infty} - T_w)$ with $\Delta T_{V\infty}$ as a parameter.

Fujii et al.[3] obtained a similarity solution for the case of a vapor with noncondensable gas using parameters of $R = 100, 500, 1000$, $Pr_L = 1, 5, 10$, and Pr_V, or $Sc = 0.2, 0.5, 1, 1.5$. They correlated the data obtained in the form of Sh_x, Nu_{wx}, and Nu_{cx}, some of which have been referred to in Chapter 3, and then proposed an algebraic method to solve the relevant problem. Later, Rose[4] formulated an accurate equation representing the

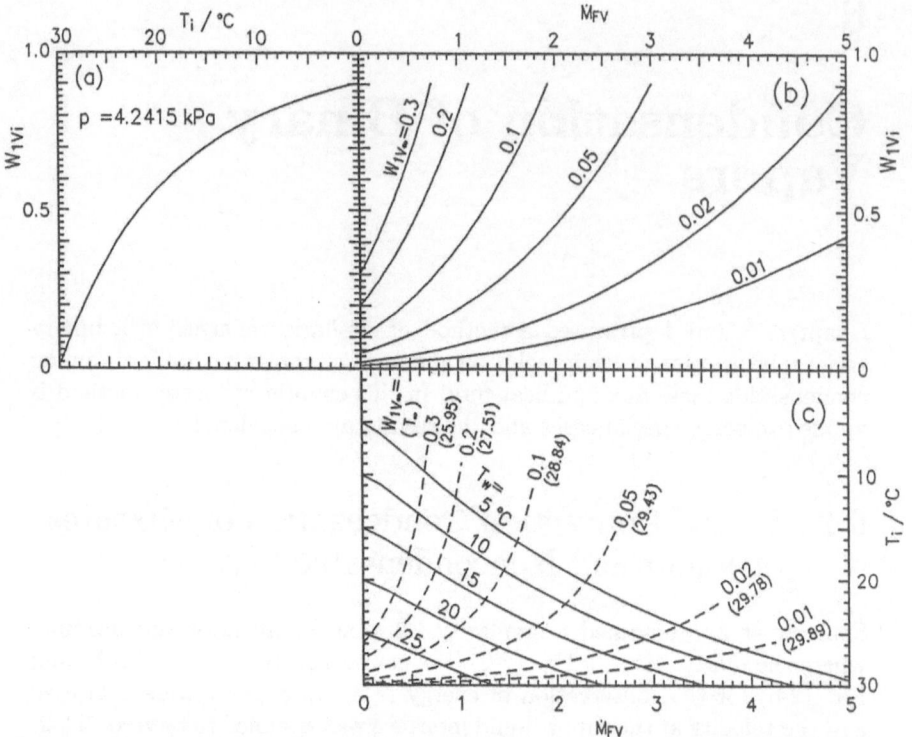

FIGURE 6.1-1. Graphical method for solving the algebraic equations of a mixture of air and saturated steam in forced-convection condensation ($p = 4.2415$ kPa). (a) W_{1Vi} versus T_i from Eqs. (6.1-4) and (2-13a), (b) $\dot{M}_{FV}$ versus W_{1Vi} from Eq. (6.1-2), (c) $\dot{M}_{FV}$ versus

relation between $Sh_x Re_{Lx}^{-1/2}$ and W_∞/W_i by using the numerical results of Fujii et al. and Koh.[14]

6.1.1 MIXTURE OF AIR AND SATURATED STEAM — GRAPHICAL SOLUTION

In this case, by neglecting the convection term in the condensate film and the convective heat transfer in the vapor phase and estimating $C_F(Sc)$ and m in the range of $R = 170$–200 and $Sc = 0.5$–0.55, Eqs. (3.6-1) and (3.6-2) are reduced to

$$0.433\left(1.367 - \frac{0.432}{\sqrt{2\dot{M}_{FV}}} + \frac{1}{2\dot{M}_{FV}}\right)^{\frac{1}{2}} = \frac{Pr_L \dot{M}_{FV}}{RH_i}, \qquad (6.1\text{-}1)$$

$$\dot{M}_{FV} = 0.331 Sc^{-0.637}\left(\frac{2.5}{1.5 + W_R}\right)^{0.51}(W_R - 1), \qquad (6.1\text{-}2)$$

where

$$W_R = \frac{W_{1Vi}}{W_{1V\infty}} \tag{6.1-3}$$

because $W_{1L}=0$. The temperature T_i at the vapor-liquid interface is equal to the saturation temperature T_s corresponding to the partial pressure of the vapor. The value T_s can be evaluated from the formula of vapor pressure, which is expressed in the form as

$$p_s = Func.(T_s) \tag{6.1-4}$$

and the relation between p_{2s} and W_{2Vi} at the vapor-liquid interface can be obtained from Eq. (2-12) or Eq. (2-13).

Solving the simultaneous algebraic equations (6.1-1), (6.1-2), (6.1-4) and (2-12) or (2-13) for given values of p, $W_{1V\infty}$, and T_w, we can obtain T_i, W_{1Vi}, and $\dot{M}_{FV}$. To reveal the simultaneity of these equations, a graphical method of solution is presented below for an air-steam mixture as an example.

Figure 6.1-1(a) shows the relation between T_i and W_{1Vi} for a mixture of air and saturated steam at $p = 4.2415$ kPa. The curve in the figure is obtained by the way that we compute $p_s = p_{2i}$ from $T_i = T_s$ using Eq. (6.1-4) and then compute W_{1Vi} from p_{2i} using Eq. (2-13a). As a matter of course, the curve depends on p. Figure 6.1-1(b) shows the relation between $\dot{M}_{FV}$ and W_{1Vi} with $W_{1V\infty}$ as a parameter in Eq. (6.1-2), where the value Sc has been evaluated at $(W_{1Vi} + W_{1V\infty})/2$ and the corresponding saturation temperature. Figure 6.1-1(c) shows the relation between T_i and $\dot{M}_{FV}$ with T_w as a parameter in Eq. (6.1-1), where the values μ_L and λ_L have been evaluated at $T_w + (T_i - T_w)/3$ and Δh_V at T_i.

We can draw the curves for the relation between $\dot{M}_{FV}$ and T_i which satisfies Eqs. (6.1-4), (2-13a), and (6.1-2) as shown by the broken lines in Fig. 6.1-1(c), using the curve in Fig. 6.1-1(a) and the curves in Fig. 6.1-1(b). The cross point of a solid line and a broken line in Fig. 6.1-1(c) is the solution of the relevant simultaneous algebraic equations. For example, we obtain $\dot{M}_{FV} = 3.25$ and $T_i = 24.1°C$ for $T_w = 5°C$ and $W_{1V\infty} = 0.02$. The corresponding algebraic solution (for $T_w = 5°C$) which is obtained by using R at $(W_{1V\infty}+W_{1Vi})/2$ and the corresponding saturation temperature is shown by the symbol $\bigcirc$ ($\dot{M}_{FV} = 3.29$, $T_i = 23.36°C$) in the figure. Generally, this graphic solution predicts a somewhat smaller value of $\dot{M}_{FV}$ and a larger value of T_i owing to inadequate use of physical properties.

We can obtain from Fig. 6.1-1 the relations between $\dot{M}_{FV}$ and $W_{1V\infty}$ and $\{T_i-T_w)/(T_{V\infty}-T_w)$ and $W_{1V\infty}$ with a parameter T_w, as shown in Fig. 6.1-2(a) and (b), respectively. These figures reveal that a small amount of air remarkably reduces the values of $\dot{M}_{FV}$ and T_i, and the decrease of $\dot{M}_{FV}$ is marked when $(T_{V\infty} - T_w)$ is large while its effect on T_i is relatively small.

The value of $\dot{m}_x$ can be obtained from Eq. (3.6-3) and $q_{wx} = \dot{m}_x \Delta h_V$ is valid because the convection term in the condensate film and the convec-

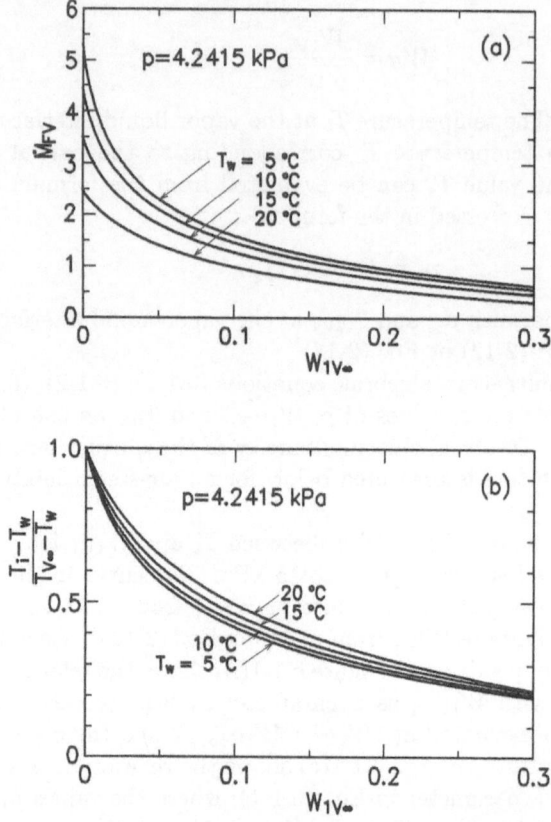

FIGURE 6.1-2. Forced-convection condensation of a mixture of air and saturated steam ($p = 4.2415$ kPa). (a) $\dot{M}_{FV}$ versus $W_{1V\infty}$. (b) $(T_i - T_w)/(T_{V\infty} - T_w)$ versus $W_{1V\infty}$.

tive heat transfer in the vapor phase are neglected. Consequently, Nu_{Lwx} corresponding to q_{wx}, which is based on $(T_i - T_w)$, is expressed as

$$Nu_{Lwx} = \frac{Pr_L}{RH} \dot{M}_{FV} Re_{Lx}^{\frac{1}{2}} . \tag{6.1-5}$$

6.1.2 NECESSARY CONDITION FOR THE APPEARANCE OF MIST AND ITS EFFECT ON CONDENSATION CHARACTERISTICS

In Section 3.3.3 we have shown that the distribution of temperature T_V, which is calculated from the energy equation, does not coincide with the saturation temperature T_{Vs} obtained from the concentration W_{1V}, and a subcool state appears, in particular, in the case of an air-water mixture. In fact, it is considered that T_V becomes T_{Vs} due to the appearance of mist or fog instead of the subcool state. The appearance of mist was treated in the research of withdrawal of toxic or precious substances, and the effect of the existence of small particles such as the condensation nucleus was discussed. However, the mechanism is not completely clarified yet.[5]

In the study of the cooling of an air-steam mixture flowing through between parallel plates, Hayashi et al.[6] experimentally confirmed that the condition of mist appearance is expressed as

$$\left[\frac{\partial (T_V - T_{Vs})}{\partial y} \right]_i < 0 . \tag{6.1-6}$$

To apply this condition to the case of condensation on a single flat surface, we derive the formula of $\partial T_s / \partial y$ first. The differentiation of Eq. (2-13a) with respect to y yields

$$\frac{\partial W_{1V}}{\partial y} = -\frac{W_{2V}}{(p - kp_2)} \frac{p}{p_2} \frac{\partial p_2}{\partial y} , \tag{6.1-7}$$

where

$$k = 1 - \frac{M_2}{M_1} . \tag{6.1-8}$$

On the other hand, the Clausius-Clapeyron's equation provides

$$\frac{\partial T_{Vs}}{\partial y} = \frac{R_2 T_{Vs}^2}{p_2 \Delta h_V} \frac{\partial p_2}{\partial y} , \tag{6.1-9}$$

where R_2 is the gas constant of steam. Eliminating $\partial p_2 / \partial y$ from Eqs. (6.1-7) and (6.1-9), we obtain

$$\frac{\partial T_{Vs}}{\partial y} = -\left(1 - k\frac{p_2}{p} \right) \frac{R_2 T_{Vs}^2}{W_{2V} \Delta h_V} \frac{\partial W_{1V}}{\partial y} . \tag{6.1-10}$$

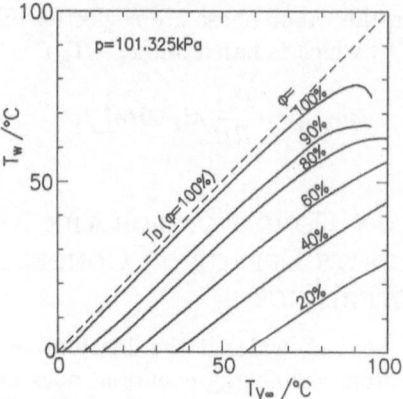

FIGURE 6.1-3. Upper limit of T_w for the appearance of subcooling in the vapor boundary layer of forced-convection condensation of an air-steam mixture at atmospheric pressure. ϕ: relative humidity.

Substituting $\partial T_V/\partial y$ from Eq. (3.1-8) and $\partial T_s/\partial y$ as a function of Φ' obtained from Eqs. (3.1-9) and (6.1-10) into Eq. (6.1-6), we obtain

$$\frac{\Delta h_V(1 - W_{1Vi})(T_{V\infty} - T_i)}{\left(1 - k\dfrac{p_{2i}}{p}\right) R_2 T_i{}^2(W_{1Vi} - W_{1V\infty})} < \frac{-\Phi'_{Fi}}{-\Theta'_{FVi}}. \tag{6.1-11}$$

This is a necessary condition of mist appearance in the present case, where the values of $-\Phi'_{Fi}$ and $-\Theta'_{FVi}$ are given by Eqs. (3.5-6) and (3.4-16), respectively. By using Eq. (6.1-11), Nagata[7] presented such a graph as shown in Fig. 6.1-3 concerning the relation between T_w and $T_{V\infty}$ with the relative humidity ϕ as a parameter for the case of an air-water mixture at atmospheric pressure. The figure means that there is a possibility of the appearance of mist in the region under respective curves.

Hijikata and Mori[8] made an analysis of the vapor boundary layer containing mist by using an integral method, and showed that the heat transfer coefficient in this case is smaller than in the case without mist. Although they assumed that the mist distributes throughout the vapor boundary layer, the subcooled region predicted from the similarity solution in Chapter 3 is confined in a narrow region near the vapor-liquid interface. Also, their heat transfer coefficient for the case without mist is smaller than that for the similarity solution by about ten percent. If the diffusion coefficient D in the vapor phase is kept constant even in the appearance of mist, T_{Vs} probably decreases according to the decrease of W_{2V}. On the other hand, since the apparent c_p of vapor containing mist is esteemed to increase, apparent Pr_V will increase, and consequently T_V should increase if the energy equation had been solved by using these apparent values. Therefore, actual temperature distribution seems to be between T_{Vs} and T_V. Clarification of this problem will need more accurate theoretical and experimental studies.

6.1.3 AN APPROXIMATE SOLUTION IN THE CASE OF SMALL VAPOR CONCENTRATION

In the case of a small value of concentration of condensing vapor in the bulk as in a humid air, the condensation mass flux becomes small and the contribution of convective heat transfer in the vapor phase to the total heat transferred becomes relatively large. In this case we can predict fairly accurate condensation characteristics by means of the following simplified method.

The convective heat transfer in the vapor boundary layer is expressed by the following equation, which corresponds to the case of $\dot{M}_{FV} \to 0$ in Eq. (3.4-14), in which $C_F(Pr_V)$ is approximated by Eq. (3.4-12b),

$$Nu_{cx} = \frac{\alpha_{cx} x}{\lambda_V} = 0.331 Pr_V^{0.363} Re_{Vx}^{\frac{1}{2}} . \qquad (6.1\text{-}12)$$

The condensation mass flux is expressed by the following equation with a similar form as the above equation:

$$Sh_x = \frac{\beta_x x}{\rho_V D} = 0.331 Sc^{0.363} Re_{Vx}^{\frac{1}{2}} . \qquad (6.1\text{-}13)$$

The heat transfer in the condensate film is expressed by the following equation[3], which is more simplified than Eq. (5.1-4),

$$Nu_{Lwx} = \frac{\alpha_{wx} x}{\lambda_L} = 0.409 \left(1.29 + \frac{1}{2\dot{M}_{FV}}\right)^{\frac{1}{2}} Re_{Lx}^{\frac{1}{2}} . \qquad (6.1\text{-}14)$$

We can derive the equations of q_{cx}, $\dot{m}_x$, and q_{wx} from Eqs. (6.1-12), (6.1-13), and (6.1-14), respectively, as

$$q_{cx} = 0.331 Pr_V^{0.363} Re_{Vx}^{\frac{1}{2}} \frac{\lambda_V}{x} (T_{V\infty} - T_i) , \qquad (6.1\text{-}15)$$

$$\dot{m}_x = 0.331 Sc^{0.363} Re_{Vx}^{\frac{1}{2}} \frac{\rho_V D}{x} \frac{(W_{1Vi} - W_{1V\infty})}{W_{1Vi}} , \qquad (6.1\text{-}16)$$

$$q_{wx} = 0.409 \left\{1.29 + \frac{Sc^{0.637} W_{1Vi}}{0.664(W_{1Vi} - W_{1V\infty})}\right\}^{\frac{1}{2}}$$
$$\times Re_{Vx}^{\frac{1}{2}} \left(\frac{\nu_V}{\nu_L}\right)^{\frac{1}{2}} \frac{\lambda_L}{x} (T_i - T_w) . \qquad (6.1\text{-}17)$$

Substituting these three equations into the equation

$$q_{cx} + \dot{m}_x \Delta h_V = q_{wx} \qquad (6.1\text{-}18)$$

we obtain the relation between W_{1Vi} and T_i. The simultaneous solution of the equation thus obtained [Eqs. (2-13a) and (6.1-4)] gives W_{1Vi} and T_i. In the particular case of $T_i \approx T_w$, without Eq. (6.1-17) q_{cx} and $\dot{m}_x$ are obtained from Eqs. (6.1-15), (6.1-16), (6.1-4), and (2-13a).

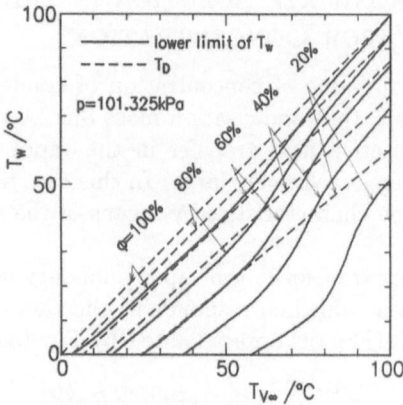

FIGURE 6.1-4. Applicable range of the approximate solution in the case of forced-convection condensation of steam from humid air at atmospheric pressure. ϕ: relative humidity.

In the case where the above approximate method is applied to the case of an air-steam mixture at atmospheric pressure, Nagata[7] presented a graph of Fig. 6.1-4, in which the lower limit of T_w, as an allowance within an error of five percent of the algebraic solution, is shown against $T_{V\infty}$ with relative humidity ϕ as a parameter. In the figure the dew points T_D are also shown by broken lines. In the region of T_w between solid and broken lines for given values of ϕ, the present approximate solution is applicable.

6.2   Free-Convection Condensation of Mixtures of Vapor and Noncondensable Gas

Sparrow and Eckert[9] obtained the state at the vapor-liquid interface and wall heat flux for free-convection condensation of steam with a small amount of air. In their analysis the buoyancy force in the vapor phase is neglected. Sparrow and Lin[10] obtained a numerical solution for the above problem, where the convection and inertia terms in the condensate film are neglected and only concentration distribution is taken into account for the buoyancy force in the vapor boundary layer. The results are presented in a graph of W_{1Vi} and/or p_{1i}/p versus H_i/Pr_L.

Minkowycz and Sparrow[11] discussed the effects of noncondensable gas in the bulk, temperature gap at the vapor-liquid interface, superheating of vapor, free-convection due to the distributions of concentration and temperature in the vapor phase, variation of physical properties of the condensate, diffusion thermo, and thermal diffusion upon the heat transfer coefficient in free-convection condensation of an air-steam mixture. However, they neglected the terms of inertia and convection in the condensate film and the shear stress at the vapor-liquid interface in the basic equations.

Fujii et al.[12] obtained a similarity solution for the case of a vapor with noncondensable gas using parameters of $R = 500$, $M_1/M_2 = 1.1$; $R = 100, 500, 1000$, $M_1/M_2 = 1.607$; and $R = 100, 500, 1000$, $M_1/M_2 = 10$; $Pr_L = 1, 5, 10$; Sc or $Pr_V = 0.2, 0.5, 1$; and $W_{1V\infty} = 0.01, 0.05, 0.1$. They correlated the obtained data in the form of Sh_x, Nu_{Lwx}, and Nu_{cx}, and proposed an algebraic method to solve the relevant problem similarly as in the case of forced-convection condensation.

6.2.1   THE CASE OF A SATURATED VAPOR AND NEGLIGIBLE CONVECTIVE HEAT TRANSFER IN THE VAPOR PHASE

In this case, as in forced-convection condensation, Eqs. (4.6-1) and (4.6-2) are simplified as

$$\dot{M}_{GL} = \left(\frac{H_i}{Pr_L}\right)^{\frac{3}{4}}, \tag{6.2-1}$$

$$\dot{M}_{GL} = \frac{2C_G(Sc)(\chi_i Sc)^{\frac{1}{4}}}{ScR} \frac{W_R - 1}{(W_R + 1)^{0.5}W_R^{1-n}}. \tag{6.2-2}$$

The simultaneous solution of these equations and Eq. (6.1-4) for given values of $T_{Vs\infty}$, $W_{1V\infty}$, and T_w yields T_i, W_{1Vi}, and $\dot{M}_{GL}$.

An example of a graphical solution is shown in Figs. 6.2-1(a),(b), and (c), which correspond to Eqs. (6.1-4), (6.2-2), and (6.2-1), respectively. The situation treated, coordinate system and graphics in these figures are the same as those in Figs. 6.1-1(a), (b), and (c), except that μ_L and λ_L in Eq. (6.2-1) have been evaluated at $T_w + (T_i - T_w)/4$. In Fig. 6.2-1(b) the fact that $\dot{M}_{GL}$ decreases with the increase of W_{1Vi} at $W_{1Vi} \approx 0.8$–0.9 stems from the large increasing rate of R in the denominator of Eq. (6.2-2), and its effect also appears in the broken lines of Fig. 6.2-1 (c). For example, the solution for $T_w = 5°C$ and $W_{1V\infty} = 0.02$ is obtained from the figure as $\dot{M}_{GL} = 1.57 \times 10^{-3}$ and $T_i = 6.0°C$. The corresponding values of the algebraic solution are $\dot{M}_{GL} = 1.526 \times 10^{-3}$ and $T_i = 6.136°C$.

The relations between $\dot{M}_{GL}$ and $W_{1V\infty}$ and $(T_i - T_w)/(T_{V\infty} - T_w)$ and $W_{1V\infty}$ with T_w as a parameter, which are obtained from Fig. 6.2-1, are shown in Figs. 6.2-2(a) and (b). When these figures are compared with Figs. 6.1-2(a) and (b) for forced-convection condensation, the decrease of $\dot{M}_{GL}$ and T_i due to the presence of air is much more drastic, particularly in the region $W_{1V\infty} = 0$–0.0005.

The values of $\dot{m}_x$ and q_{wx} are evaluated by Eqs. (4.6-3) and (4.6-4), respectively, and Nu_{Lwx} is evaluated by Eqs. (4.4-1) and (4.4-4).

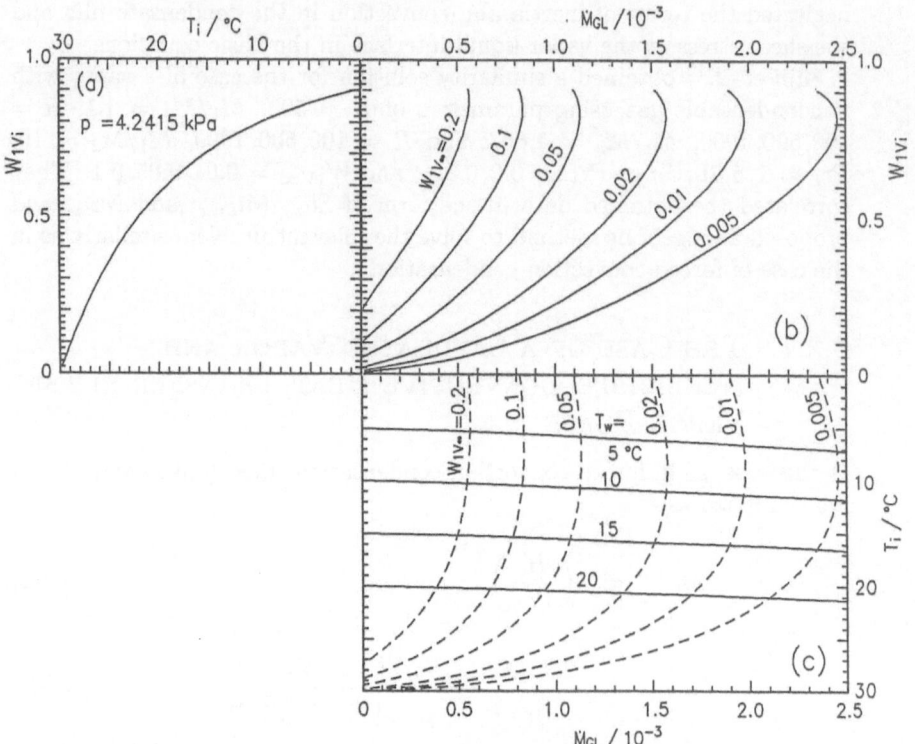

FIGURE 6.2-1. Graphical method for solving the algebraic equations of a mixture of air and saturated steam in free-convection condensation ($p = 4.2415$ kPa). (a) W_{1Vi} versus T_i from Eq. (6.1-4), (b) $\dot{M}_{GL}$ versus W_{1Vi} from Eq. (6.2-2), (c) $\dot{M}_{GL}$ versus T_i from Eq. (6.2-1).

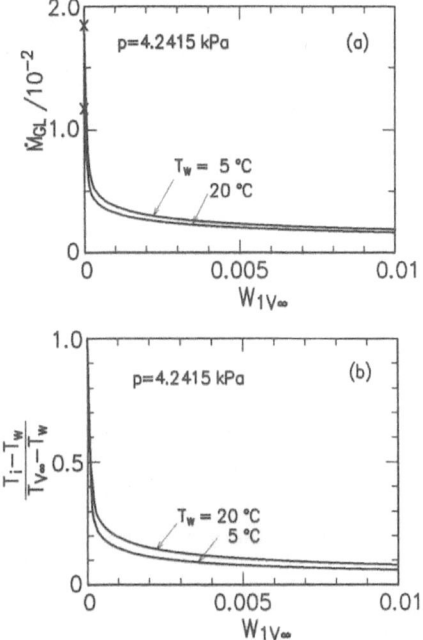

FIGURE 6.2-2. Free-convection condensation of a mixture of air and saturated steam (p=4.2415 kPa). (a) $\dot{M}_{GL}$ versus $W_{1V\infty}$, (b) $(T_i - T_w)/(T_{V\infty} - T_w)$ versus $W_{1V\infty}$.

6.2.2   SINGLE PHASE FREE-CONVECTION WITH SIMULTANEOUS HEAT AND MASS TRANSFER —THE CASE OF VERY SMALL CONDENSATION MASS FLUX

The equations relevant to this case can be derived from Eqs. (4.4-5) and (4.5-2), which are contained in Eqs. (4.6-1) and (4.6-2), respectively, as

$$\sqrt{2}Nu_x(Gr_x)^{-\frac{1}{4}} = \sqrt{2}C_G(Pr)\left(\frac{\chi_i^+ Pr}{\omega}\right)^{\frac{1}{4}}, \qquad (6.2\text{-}3)$$

$$\sqrt{2}Sh_x(Gr_x)^{-\frac{1}{4}} = \sqrt{2}C_G(Sc)\left(\frac{\chi_i Sc}{\omega}\right)^{\frac{1}{4}}, \qquad (6.2\text{-}4)$$

where

$$Gr_x = \frac{x^3 g\omega}{\nu_V^2}, \qquad (6.2\text{-}5)$$

$$\omega = \omega_W + \omega_T. \qquad (6.2\text{-}6)$$

Gebhart and Pera[13] presented a similarity solution of this problem in the ranges of $Pr = 0.7$ and 7, $Sc = 0.1$–500, and $\omega_W/\omega_T = -0.5$–2. The

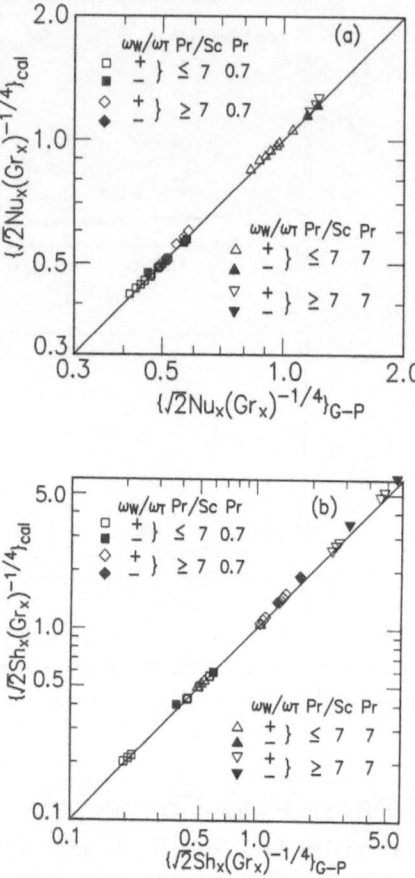

FIGURE 6.2-3. Comparison between Gebhart and Pera's similarity solution (subscript G–P) and the present correlation equations (subscript cal) for simultaneous heat and mass transfer in single phase free convection. (a) $[\sqrt{2}Nu_x(Gr_x)^{-1/4}]_{cal}$ [Eq. (6.2-3)] versus $[\sqrt{2}Nu_x(Gr_x)^{-1/4}]_{G-P}$, (b) $[\sqrt{2}Sh_x(Gr_x)^{-1/4}]_{cal}$ [Eq. (6.2-4)] versus $[\sqrt{2}Sh_x(Gr_x)^{-1/4}]_{G-P}$.

boundary values of their solution for heat and mass transfer correspond to the values on the left-hand side of Eqs. (6.2-3) and (6.2-4), respectively. χ_i^+/ω and χ_i/ω on the right-hand side of Eqs. (6.2-3) and (6.2-4) can be approximated by the function of ω_W/ω_T. Figures 6.2-3(a) and (b) show the comparison of Eqs. (6.2-3) and (6.2-4) with the corresponding values of Gebhart and Pera, respectively. We can see that these equations correlate well with the numerical solution.

6.3 Forced-Convection Condensation of Binary Vapors

Koh[14] derived the two-phase boundary layer equations for the laminar film forced-convection condensation of a binary vapor mixture and established the method to solve them by means of the similarity transformation. In his analysis, however, the diffusion term in Eq. (2-6) is neglected. He presented the numerical results for $R = 10, 100, 500$, $Pr_L = 0.003, 0.008, 0.03, 1, 10, 100$, $Sc = Pr_V = 1, 5$, and $\eta_{Li} = 0.1\text{--}3.5$ in a table, and showed examples for condensation mass flux, shear force at the cooling surface, and heat transfer coefficient of a saturated pure vapor, a superheated pure vapor, a binary vapor mixture, and a vapor with noncondensable gas by using the numerical table. Fujii et al.[15] presented a detailed numerical result on the relevant problem and proposed some correlation equations. Some of them have been referred to in Chapter 3.

6.3.1 GRAPHICAL SOLUTION AND SOME TYPICAL EXAMPLES

Figure 6.3-1 shows a graphical method of solution for forced-convection condensation of a saturated methanol-water vapor mixture at $p = 0.1$ MPa ($T_s = 89.46°C$) as an example. Figure 6.3-1(a) shows a phase equilibrium diagram corresponding to Eqs. (2-29) and (2-30). Figure 6.3-1(b) shows the relation between W_R and $T_i(W_{1Vi}, W_{1L})$ with $W_{1V\infty}$ as a parameter in Eq. (3.1-37). Figure 6.3-1(c) shows the relation between $\dot{M}_{FV}$ and W_R in Eq. (3.6-2), where Sc is evaluated at $(W_{1V\infty} + W_{1Vi})/2$ and corresponding saturation temperature. Figure 6.3-1(d) shows the relation between $\dot{M}_{FV}$ and T_i with T_w as a parameter in Eq. (3.6-1), where the convective heat transfer in the vapor boundary layer is neglected and physical properties in the condensate film are evaluated at $T_w + (T_i - T_w)/3$ and R is evaluated at T_i. The broken lines in Fig. 6.3-1(d) represent the relation between $\dot{M}_{FV}$ and T_i which is obtained from Figs. 6.3-1(b) and (c). The cross points between the solid and broken lines correspond to the solution; e.g., we can obtain $\dot{M}_{FV} = 1.09$, $T_i = 78.0°C$, $W_{1Vi} = 0.776$, and $W_{1L} = 0.415$

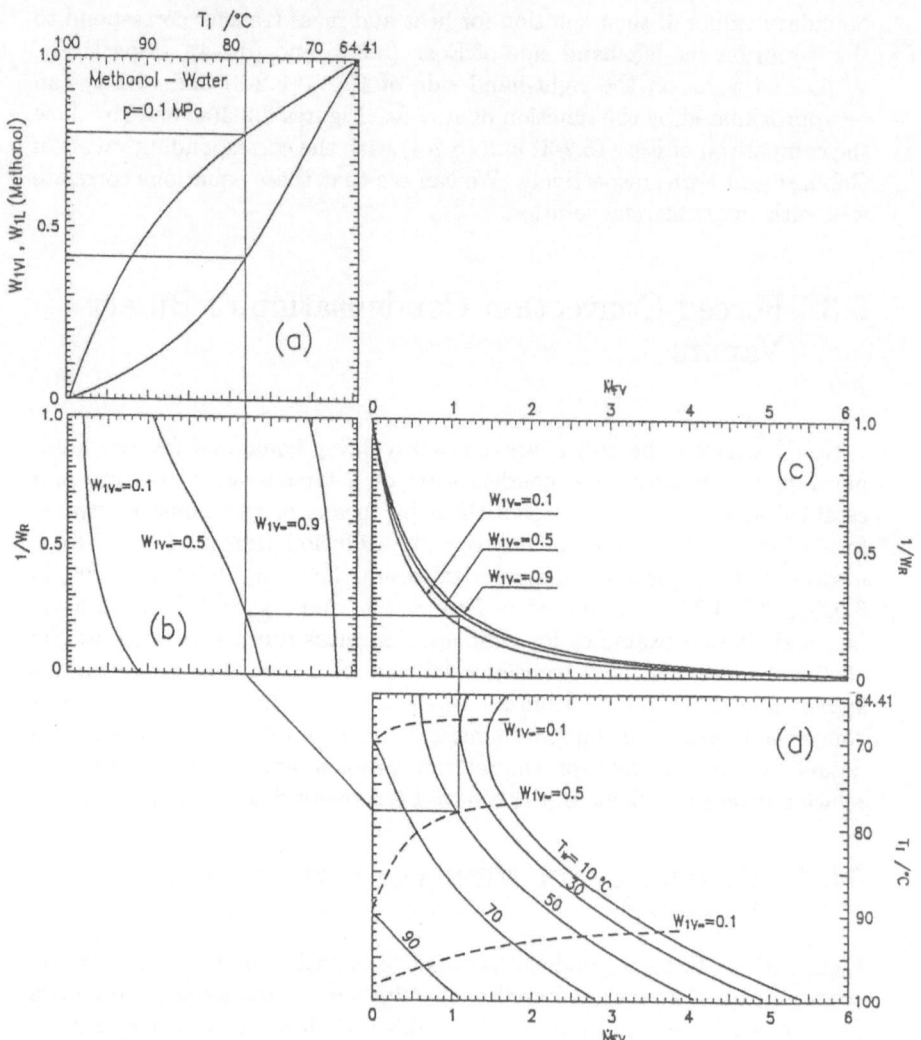

FIGURE 6.3-1.Graphical method for solving the algebraic equations of an methanol-water mixture in forced-convection condensation (p= 0.1 MPa). (a) Phase equilibrium, (b) Eq. (3.1-37), (c) Eq. (3.6-2), (d) Eq. (3.6-1).

TABLE 6.3-1. Comparison among the similarity, algebraic, and stagnant film theories' solutions for a methanol-water mixture at $p=0.1$ MPa, $W_{1V\infty} = 0.5$, and $T_w = 50°C$.

	$\dot{M}_{FV}$	$\dfrac{T_i}{°C}$	W_{1Vi}	W_{1L}	$\dot{m}_x(x/U_{V\infty})^{1/2}$ kgm^{-2}s$^{-1/2}$ × 10^{-3}	$q_{wx}(x/U_{V\infty})^{1/2}$ Wm^{-2}s$^{1/2}$ × 10^3	$q_{cx}(x/U_{V\infty})^{1/2}$ Wm^{-2}s$^{1/2}$ × 10^3
similarity solution	1.0929	78.01	0.7760	0.4155	3.425	6.350	0.083
algebraic solution	1.0861	78.03	0.7756	0.4147	3.403	6.362	0.083
stagnant film theory	1.0386	76.99	0.7942	0.4549	3.264	5.793	0.079

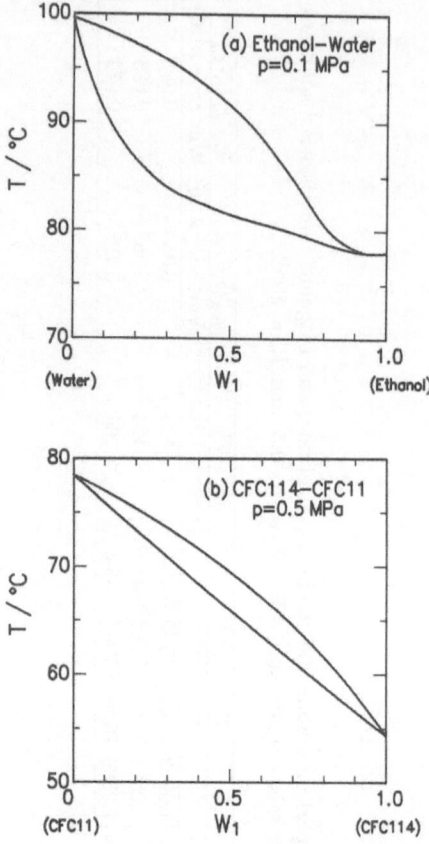

FIGURE 6.3-2. Examples of the phase equilibrium diagram for binary vapor mixtures. (a) Ethanol-water, (b) CFC114-CFC11.

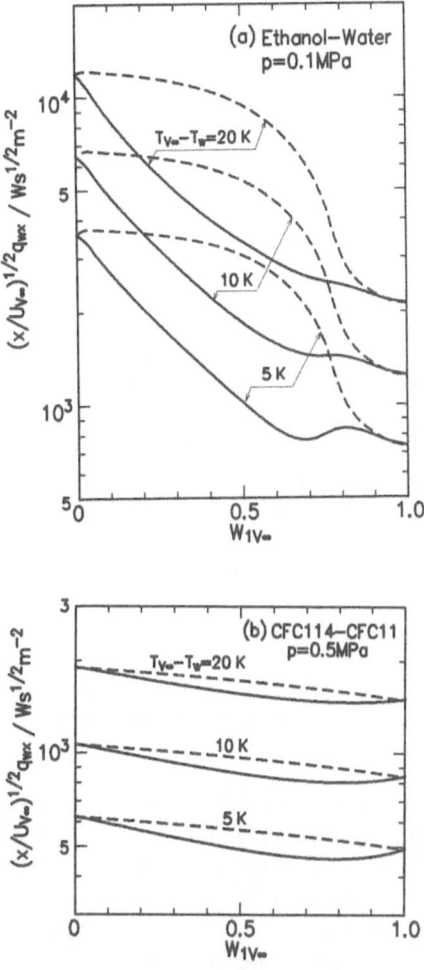

FIGURE 6.3-3.   Examples of the relation between $q_{wx}(x/U_{V\infty})^{1/2}$ and $W_{1V\infty}$ in forced-convection condensation of binary vapor mixtures. (a) Ethanol-water, (b) CFC114-CFC11.

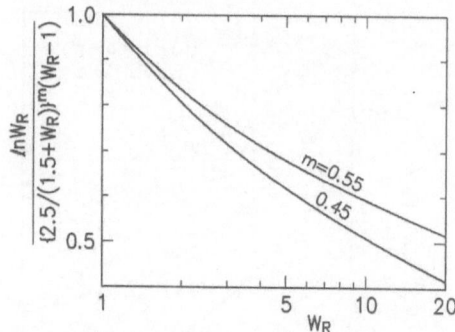

FIGURE 6.3-4. Ratio of $(\dot{M}_{FV})_{film}$ of Eq. (6.3-2) for the stagnant film theory to $(\dot{M}_{FV})_{cal}$ of Eq. (3.6-2) for the similarity solution.

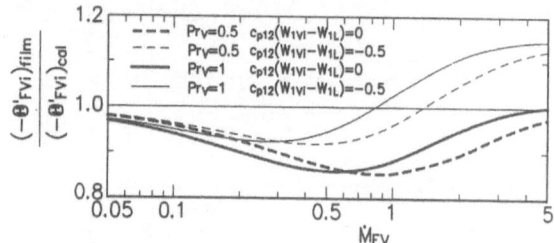

FIGURE 6.3-5. Ratio of $(-\Theta'_{FVi})_{film}$ of Eq. (6.3-3) for the stagnant film theory to $(-\Theta'_{FVi})_{cal}$ of Eq. (3.4-16) for the similarity solution.

for a given condition of $p = 0.1\text{MPa}$, $W_{1V\infty} = 0.5$, and $T_w = 50°\text{C}$. The corresponding algebraic solution is shown in Table 6.3-1.

Figures 6.3-2(a) and (b) show the phase equilibrium diagrams of an ethanol-water mixture at 0.1 MPa and a CFC114-CFC11 mixture at 0.5 MPa, respectively. The former is a typical example of a mixture in which the temperature difference between the dew point and boiling point is large and an azeotropic point exists. The latter is a typical example of a mixture in which the temperature difference is small and which can be esteemed as an ideal mixture. We now compare these two mixtures in the characteristics of forced-convection condensation.

Figures 6.3-3(a) and (b) show the relation between $q_{wx}(x/U_{V\infty})^{1/2}$ and $W_{1V\infty}$ with $(T_{V\infty} - T_w)$ as a parameter, where (a) and (b) correspond to Figs. 6.3-2(a) and (b), respectively. In Figs. 6.3-3(a) and (b) solid lines represent the algebraic solutions of Eqs. (3.6-1) and (3.6-2) and broken lines are obtained from Eq. (5.1-8) under the assumption that the mixture behaves like a single component vapor. The difference in q_{wx} between the solid and broken lines is large in the region $W_{1V\infty} \approx 0.5$–0.7 and when $(W_{1V\infty} - T_w)$ is small, and the difference in (a) is more remarkable than that in (b). In (a), as a matter of course, the solid and broken lines join near the azeotropic point.

6.3.2 ACCURACY OF THE STAGNANT FILM THEORY

In chemical engineering the stagnant film theory[16] is used in the analysis of condensation of a binary vapor or multicomponent vapor. We now discuss the accuracy of the theory.

The equation of the stagnant film theory corresponding to Eq. (3.5-6) is written as

$$\frac{-\Phi'_{Fi}}{C_F(Sc)} = \frac{W_R}{(W_R - 1)} \ln W_R .\tag{6.3-1}$$

From this equation we can derive

$$\dot{M}_{FV} = \frac{C_F(Sc)}{Sc} \ln W_R .\tag{6.3-2}$$

Figure 6.3-4 shows the relation between the ratio of $\dot{M}_{FV}$ in Eq. (6.3-2) to $\dot{M}_{FV}$ in Eq. (3.6-2). We can see that the ratio considerably decreases with the increase of W_R.

The equation of the stagnant film theory for the convective heat transfer in the vapor phase is written as

$$\frac{-\Theta'_{FVi}}{C_F(Pr_V)} = \frac{\dfrac{Pr_V}{C_F(Pr_V)}(\dot{M}_{FV})\{1 - c_{p12}(W_{1Vi} - W_{1L})\}}{1 - \exp\left[\dfrac{Pr_V}{C_F(Pr_V)}(-\dot{M}_{FV})\{1 - c_{p12}(W_{1Vi} - W_{1L})\}\right]} .\tag{6.3-3}$$

Figure 6.3-5 shows the ratio of $(-\Theta'_{FVi})_{film}$ of Eq. (6.3-3) to $(-\Theta'_{FVi})_{cal}$ of Eq. (3.4-16). The maximum deviation of this ratio from unity is about 15 percent, although it depends on Pr_V and $c_{p12}(W_{1Vi} - W_{1L})$.

Using Eq. (6.3-3) we can derive the equation corresponding to Eq. (3.6-1) as

$$0.433\left(1.367 - \frac{0.432}{\sqrt{2\dot{M}_{FV}}} + \frac{1}{2\dot{M}_{FV}}\right)^{\frac{1}{2}} = \frac{Pr_L \dot{M}_{FV}}{RH_i} +$$

$$\frac{\lambda_V \sqrt{\nu_L}(T_{V\infty} - T_i) Pr_V \dot{M}_{FV}\{1 - c_{p12}(W_{1Vi} - W_{1L})\}}{\lambda_V \sqrt{\nu_L}(T_i - T_w)\left\langle 1 - \exp\left[\dfrac{Pr_V}{C_F(Pr_V)}(-\dot{M}_{FV})\{1 - c_{p12}(W_{1Vi} - W_{1L})\}\right]\right\rangle} .\tag{6.3-4}$$

Table 6.3-1 shows an example of the comparison among the similarity solution, the algebraic solution using Eqs. (3.6-1) and (3.6-2), and the solution of stagnant film theory using Eqs. (6.3-2) and (6.3-4) for a methanol-water mixture at $p = 0.1$ MPa, $W_{1V\infty} = 0.5$, and $T_w = 50°C$. Although the algebraic solution agrees well with the similarity solution, the stagnant film theory predicts lower values of $\dot{m}_x$, q_{wx}, and $(T_i - T_s)$ by about five, ten, and five percent, respectively.

6.4 Free-Convection Condensation of Binary Vapors

Sparrow and Marschall[17] treated free-convection condensation of a methanol-water mixture by the same method as that of Sparrow and Lin[10], and presented the results in a graph showing the relation of $q_{wx}/(q_{wx})_{Nu}$ versus $(T_{V\infty} - T_w)$ with $T_{V\infty}$ as a parameter [where $(q_{wx})_{Nu}$ is the heat flux obtained from Eq. (5.3-2) under the assumption that the condensation takes place at $T_i = T_{V\infty}$ and $W_{1Vi} = W_{1V\infty}$]. Koyama et al.[18] presented a detailed numerical result on the relevant problem and proposed some correlation equations. Some of them have been referred to in Chapter 4.

Figures 6.4-1(a) and (b) show examples of the relations between $q_{wx}x^{1/4}$ and $W_{1V\infty}$, for an ethanol-water vapor mixture at 0.1 MPa and a CFC114-CFC11 vapor mixture at 0.5 MPa, respectively. In the figures the solid lines represent the solution of Eqs. (4.6-1) and (4.6-2), and the broken lines represent the values obtained from Eq. (5.3-2) under the assumption that the mixture behaves as a single component vapor. In the comparisons between Figs. 6.3-3(a) and 6.4-1(a), between Figs. 6.3-3(b) and 6.4-1(b), and between Figs. 6.4-1(a) and (b), we find that the tendency shown in these figures is almost the same as in forced- and free-convection condensation, except the magnitude of the difference between the solid lines and broken lines is much more in the latter.

As stated at the beginning of Chapter 4, no similarity solution exists in the case where molecular weight M_1 for the volatile component is smaller than M_2 for the less volatile component. In the case where the molecular weight of noncondensable gas is smaller than that of vapor, Hijikata et al.[19] obtained an approximate solution using an integral method, and found that the average heat transfer coefficient can be predicted from the previously established equation, in which the absolute value of buoyancy force is taken. Goto and Fujii[20] experimentally confirmed that the average wall heat flux for such a binary vapor mixture can be predicted from Eqs. (4.6-1) and (4.6-2) in which absolute values of χ_i and χ_i^+ are used.

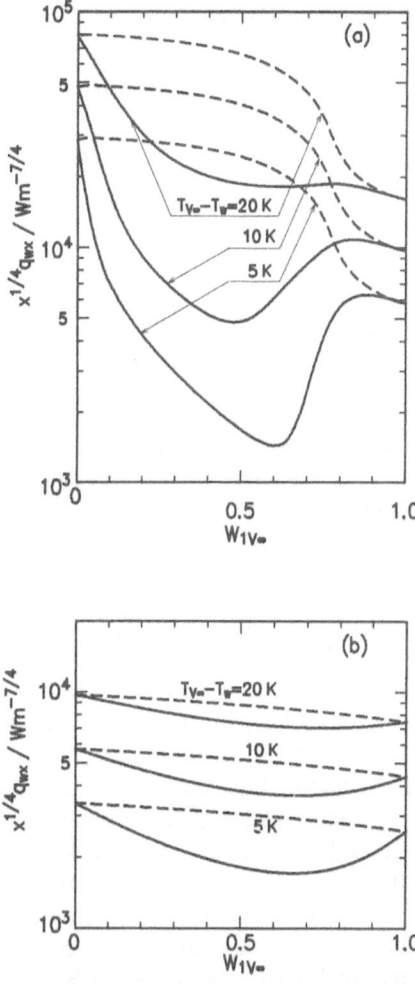

FIGURE 6.4-1.   Examples of the relation between $q_{wx}x^{\frac{1}{4}}$ and $W_{1V\infty}$ in free-convection condensation of binary vapor mixtures. (a) Ethanol-water at $p{=}0.1$ MPa, (b) CFC114-CFC11 at $p{=}0.5$ MPa.

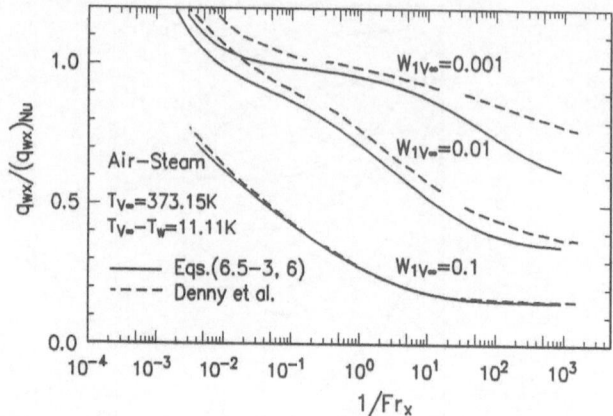

FIGURE 6.5-1. Comparison of the algebraic solution with the finite difference solution of Denny et al. for combined forced- and free-convection condensation of an air-steam mixture.

6.5 Combined Forced- and Free-Convection Condensation of Binary Vapors

In this case the similarity transformation is impossible. Accordingly, some examples of numerical solutions by means of a finite difference method are reported: for steam with a small amount of air by Denny et al.,[21] for water, ammonia, ethanol, buthanol, and carbon tetrachloride with a small amount of air by Denny and Jusionis,[22] for methanol-water, propanol-water, aceton-water, and aceton-carbontetrachloride mixtures by Denny and Jusionis,[23] and for a methanol-water mixture by Lucas.[24] Stephan and Laesecke[25] modified the film theory for simultaneous heat and mass transfer by taking account of mass flux normal to the wall, and compared their result with that of Lucas. As stated in Section 5.5, the numerical solution of combined forced- and free-convection condensation of a pure vapor can be correlated by a combination of the respective equations for forced- and free-convection condensation. Fujii and Kato[26] applied this method to the present problem and compared the result with the above numerical solution. This is explained below. For simplicity, it is assumed that the bulk vapor mixture is saturated and the effect of convection in the condensate film is neglected.

Equations (3.6-1) and (4.6-1) are rewritten as

$$\frac{Pr_L\dot{M}_{FL}}{H_i} = 0.433\left(1.367 - \frac{0.432}{\sqrt{2\dot{M}_{FV}}} + \frac{1}{2\dot{M}_{FV}}\right)^{\frac{1}{2}} = \frac{Nu_{Lwx}}{Re_{Lx}^{\frac{1}{2}}},$$
(6.5-1)

$$\frac{Pr_L\dot{M}_{GL}}{H_i} = \dot{M}_{GL}^{-\frac{1}{3}} = \frac{\sqrt{2}Nu_{Lwx}}{Ga_x^{\frac{1}{4}}} = \frac{\sqrt{2}Nu_{Lwx}}{Re_{Lx}^{\frac{1}{2}}}Fr_x^{\frac{1}{4}}.$$
(6.5-2)

The 1/4th power of the sum of the 4th powers of Eqs. (6.5-1) and (6.5-2) gives

$$\frac{Nu_{Lwx}}{\sqrt{Re_{Lx}}} = \left\{(0.433)^4\left(1.367 - \frac{0.432}{\sqrt{2R\dot{M}_{FL}}} + \frac{1}{2R\dot{M}_{FL}}\right)^2 \right.$$
$$\left. + (4\dot{M}_{FL}Fr_x)^{-\frac{4}{3}}\right\}^{\frac{1}{4}}$$
$$= \frac{Pr_L\dot{M}_{FL}}{H_i}.$$
(6.5-3)

Further, Eqs. (3.6-2) and (4.6-2) can be rewritten as

$$\frac{\dot{m}_x x}{\mu_L} = \frac{C_F(Sc)}{ScR}\left(\frac{2.5}{1.5+W_R}\right)^m(W_R - 1)Re_{Lx}^{\frac{1}{2}},$$
(6.5-4)

$$\frac{\dot{m}_x x}{\mu_L} = \frac{2C_G(Sc)}{ScR}(\chi_i Sc)^{\frac{1}{4}}\frac{(W_R - 1)}{(W_R - 1)^{0.5}W_R^{1-n}}\left(\frac{Ga_x}{4}\right)^{\frac{1}{4}}.$$
(6.5-5)

The 1/4th power of the sum of the 4th powers of Eqs. (6.5-4) and (6.5-5) gives

$$\frac{\dot{m}_x x}{\mu_L Re_{Lx}^{\frac{1}{2}}} = \left[\left\{\frac{C_F(Sc)}{ScR}\left(\frac{2.5}{1.5+W_R}\right)^m(W_R - 1)\right\}^4 \right.$$
$$\left. + \left\{\frac{2C_G(Sc)}{ScR}(\chi_i Sc)^{\frac{1}{4}}\frac{(W_R - 1)}{(W_R - 1)^{0.5}W_R^{1-n}}\right\}^4\frac{1}{4Fr_x}\right]^{\frac{1}{4}}$$
$$= \dot{M}_{FL}.$$
(6.5-6)

We can obtain $\dot{M}_{FL}$, T_i, W_{1Vi}, and W_{1L} by simultaneously solving Eqs. (6.5-3), (6.5-6), (2-29), (2-30), and (3.1-37) with Fr_x as a parameter. The result for an air-steam mixture of $T_{V\infty} = 373.15$ K and $\Delta T_{V\infty} = 11.11$ K is shown by solid lines in Fig. 6.5-1, and the result of Denny et al.[21] for the same condition is also shown by broken lines in the figure [where in the ordinate q_{wx} is the surface heat flux and $(q_{wx})_{Nu}$ is that calculated from Eq. (5.3-2)]. In this figure both results agree well for $W_{1V\infty} = 0.1$, while for small values of $W_{1V\infty}$ the values of $q_{wx}/(q_{wx})_{Nu}$ from Denny et al. are

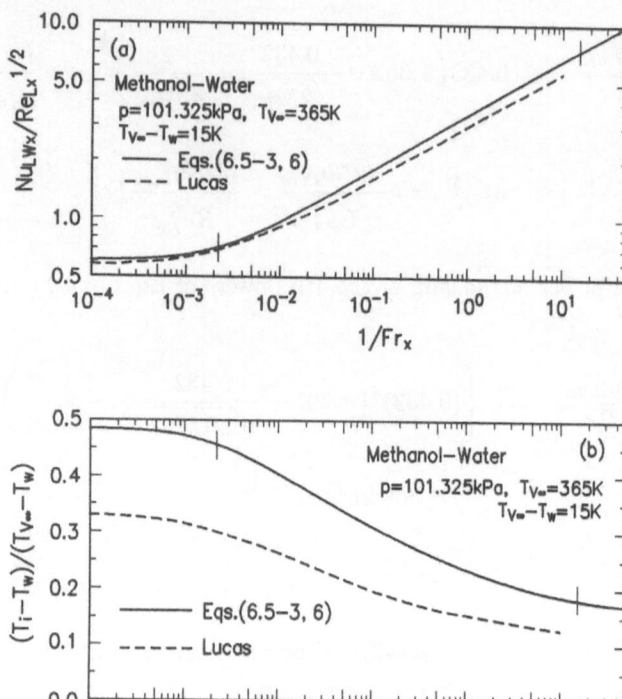

FIGURE 6.5-2.   Comparison of the algebraic solution with Lucas's finite difference solution for combined forced- and free-convection condensation of an methanol-water mixture.   (a) $Nu_{Lwx}Re_{Lx}^{-1/2}$ versus Fr_x^{-1}, (b) $(T_i - T_w)/(T_{V\infty} - T_w)$ versus Fr_x^{-1}.

somewhat larger but the tendency is similar.

Figures 6.5-2(a) and (b) compare the present solution represented by solid lines with Lucas's[24] result represented by broken lines for a methanol-water mixture of $p = 101.325$ kPa, $T_{V\infty} = 365$ K, and $T_{V\infty} - T_w = 15$ K, where (a) is for the relation of $Nu_{Lwx}Re_{Lx}^{-1/2}$ versus Fr_x^{-1} and (b) is for the relation of $(T_i - T_w)/(T_{V\infty} - T_w)$ versus Fr_x^{-1}. In both figures the tendency of solid and broken lines is similar. However, there is some difference in magnitude, which probably stems from the difference of physical properties used, because the limit values corresponding to forced-convection and free-convection condensations are different, especially in $(T_i - T_w)$ values. The vertical short lines at $Fr_x^{-1} = 2.2 \times 10^{-3}$ and $Fr_x^{-1} = 15.2$ in the figures represent the values of Fr_x at which the forced-convection term equals the free-convection term in Eqs. (6.5-3) and (6.5-6), respectively. The transition from forced-convection condensation to free-convection condensation in the relation of $Nu_{Lwx}Re_{Lx}^{-1/2}$ versus Fr_x^{-1} takes place near $Fr_x^{-1} = 2.2 \times 10^{-3}$, while $(T_i - T_w)$ varies gradually in the region of $Fr_x^{-1} = 2.2 \times 10^{-3}$ to 15.2.

REFERENCES

1. Sparrow, E. M., J. W. Minkowycz, and M. Saddy, Forced Convection Condensation in the Presence of Noncondensables and Interfacial Resistance, *Int. J. Heat Mass Transfer*, 10, 1829–1845 (1967).

2. Minkowycz, W. J. and E. M. Sparrow, The Effect of Superheating on Condensation Heat Transfer in a Forced Convection Boundary Layer Flow, *Int. J. Heat Mass Transfer*, 12, 147–154 (1969).

3. Fujii, T., H. Uehara, K. Mihara, and Y. Kato, Forced Convection Condensation in the Presence of Noncondensables. –A Theoretical Treatment for Two-Phase Laminar Boundary Layer (in Japanese), *Reports of Research Institute of Industrial Science, Kyushu University*, No. 66, 53–80 (1977). *The 2nd Report,* No. 71, 29–35 (1980).

4. Rose, J. W., Approximate Equations for Forced-Convection Condensation in the Presence of a Non-condensing Gas on a Flat Plate and Horizontal Tube, *Int. J. Heat Mass Transfer*, 23, 539–546 (1980).

5. Johnstone, H. F., Max D. Kelley, and D. L. McKinley, Fog Formation in Cooler-Condensers, *Industrial and Engineering Chemistry*, 42, 11, 2298–2302 (1950).

6. Hayashi, Y., A. Takimoto, and Y. Yamamoto, Heat and Mass Transfer with a Mist Formation in a Laminar Duct Flow , *Heat Transfer Japanese Research*, 10, 1, 37–51(1981).

7. Nagata, T., Heat and Mass Transfer of Humid Air Through a Horizontal Tube Bundle (in Japanese), *Doctoral Thesis, Kyushu University* (1982).

8. Hijikata, K. and Y. Mori, Forced Convective Heat Transfer of a Gas Containing a Condensing Vapor on a Flat Plate (in Japanese), *Trans. Jpn. Soc. Mech. Eng.*, 38, 314, 2630–2640 (1972).

9. Sparrow, E. M. and E. R. G. Eckert, Effects of Superheated Vapor and Noncondensable Gases on Laminar Film Condensation, *AIChE J*, 7, 3, 473–477 (1961).

10. Sparrow, E. M. and S. H. Lin, Condensation Heat Transfer in the Presence of a Noncondensable Gas, *Trans. ASME, J. Heat Transfer*, C, 86, 430–436 (1964).

11. Minkowycz, W. J. and E. M. Sparrow, Condensation Heat Transfer in the Presence of Noncondensables, Interfacial Resistance, Superheating, Variable Properties, and Diffusion, *Int. J. Heat Mass Transfer*, 9, 1125–1144 (1966).

12. Fujii, T. , H. Uehara, K. Mihara, and H. Takashima, Body Force Convection Condensation in the Presence of Noncondensables (in Japanese), *Reports of Research Institute of Industrial Science, Kyushu University*, No. 67, 23–41 (1978).

13. Gebhart, B. and L. Pera, The Nature of Vertical Natural Convection Flows Resulting from the Combined Buoyancy Effects of Thermal and Mass Diffusion, *Int. J. Heat Mass Transfer*, 14-12, 2025–2050 (1971).

14. Koh, J. C. Y., Laminar Film Condensation of Condensible Gases and Gaseous Mixtures on a Flat Plate, *Proc. 4th U.S. National Congress of Applied Mechanics*, 2, 1327–1336 (1962).

15. Fujii, T. ,Sh. Koyama, and M. Watabe, Laminar Forced-convection Condensation of Binary Mixtures on a Flat Plate (in Japanese), *Trans. Jpn. Soc. Mech. Eng.*, 53, 486, B, 541–548 (1987).

16. Bird. R. B., W. E. Stewart, and E. N. Lightfoot, "Transport Phenomena," John Wiley, New York (1960).

17. Sparrow, E. M. and E. Marschall, Binary, Gravity-Flow Film Condensation, *J. Heat Transfer*, 91, 205–211 (1969).

18. Koyama, Sh. , M. Watabe, and T. Fujii, The Gravity Controlled Film Condensation of Saturated and Superheated Binary Vapour Mixtures on a Vertical Plate (in Japanese), *Trans. Jpn. Soc. Mech. Eng.*, 52, 474, B, 827–834 (1986).

19. Hijikata, K. , Y. Mori, and K. Utsunomiya, Nonsimilar Solution of Condensation Heat Transfer Containing Noncondensable Gas (in Japanese), *Trans. Jpn. Soc. Mech. Eng.*, 46, 408, B, 1514–1522 (1980).

20. Goto, N. and T. Fujii, Film Condensation of Binary Refrigerant Vapors on a Horizontal Tube, *Proc. 7th Int. Heat Transfer Conf. (Munich)*, 5, LS12, 71–76 (1982).

21. Denny, V. E. , A. F. Mills, and V. J. Jusionis, Laminar Film Condensation From a Steam-Air Mixture Undergoing Forced Flow Down a Vertical Surface, *Trans. ASME, C, J. Heat Transfer*, 93-3, 297–304 (1971).

22. Denny, V. E. and V. J. Jusionis, Effects of Noncondensable Gas and Forced Flow on Laminar Film Condensation, *Int. J. Heat Mass Transfer*, 15, 315–326 (1972).

23. Denny, V. E. and V. J. Jusionis, Effects of Forced Flow and Variable Properties of Binary Film Condensation, *Int. J. Heat Mass Transfer*, 15, 2143–2153 (1972).

24. Lucas, K., Combined Body Force and Forced Convection in Laminar Film Condensation of Mixed Vapours–Integral and Finite Difference Treatment, *Int. J. Heat Mass Transfer*, 19-11, 1273–1280 (1976).

25. Stephan, K. and A. Laesecke, The Influence of Suction on Heat and Mass Transfer in Condensation of Mixed Vapors, *Wärme-und Stoffübertragung*, 13, 115–123 (1980).

26. Fujii, T. and Y. Kato, Laminar Film Condensation of a Binary Vapour on a Flat Surface (in Japanese), *Trans. Jpn Soc. Mech. Eng.*, 46, 402, B, 306–312 (1980).

7

Forced-Convection Condensation of Multicomponent Vapors

Toor[1] and Stewart and Prober[2] derived the boundary layer equations for a multicomponent vapor mixture by linearizing the relevant diffusion coefficient. They then transformed the equations to the same form as those in a binary vapor mixture by using a matrix method. Toor discussed that the solution for forced-convection mass transfer of a binary vapor mixture can be extended to the case of the multicomponent mixture. Stewart and Prober obtained a solution for forced-convection condensation of a hydrogen-nitrogen-carbon dioxide mixture by combining the matrix method and stagnant film theory, and showed that the agreement between the result and corresponding solution of the boundary layer equation is good. Fujii and Koyama[3] proposed an algebraic method for solving the condensation problem of a ternary vapor mixture, which is derived by a matrix transformation of ordinary differential equations using the result of a binary vapor mixture case. Koyama et al.[4] successfully extended this method to a multicomponent condensation problem.

7.1 Basic Equations for Forced-Convection Condensation of a Multicomponent Vapor

The same physical model as in Chapter 2 is valid when W_{1L} and W_{1V} are replaced by W_{kL} and $W_{kV}(k = 1,\ldots,n-1)$ in Fig. 2-1. Accordingly, Eqs. (2-6) and (2-7) are replaced, respectively, by

$$U_V \frac{\partial T_V}{\partial x} + V_V \frac{\partial T_V}{\partial y} = \kappa_V \frac{\partial^2 T_V}{\partial y^2} + \frac{\partial T_V}{\partial y} \sum_{k=1}^{n-1} \left(\sum_{l=1}^{n-1} c_{pkn} D_{kl}^+ \frac{\partial W_{lV}}{\partial y} \right),$$

$$(7.1\text{-}1)$$

$$U_V \frac{\partial W_{kV}}{\partial x} + V_V \frac{\partial W_{kV}}{\partial y} = \sum_{l=1}^{n-1} D_{kl}^+ \frac{\partial^2 W_{lV}}{\partial y^2} \quad (k = 1, 2, \ldots, n-1),$$

$$(7.1\text{-}2)$$

where

$$c_{pkn} = \frac{c_{pk} - c_{pn}}{c_p} \quad (k = 1, 2, \ldots, n - 1), \tag{7.1-3}$$

$$D_{kl}^+ = D_{kn} - D_{kl} \quad (k, l = 1, 2, \ldots, n - 1), \tag{7.1-4}$$

$$D_{kl} = M_k D_{kl}^M \sum_{m=1}^{n} \frac{W_{mV}}{M_m} - \frac{M_k}{M_l} \sum_{m=1}^{n} D_{km}^+ W_{mV}$$
$$(k, l = 1, 2, \ldots, n - 1) \tag{7.1-5}$$

and D_{kl}^M is the diffusivity of the pair of components $k-\ell$ in a multicomponent mixture, the value of which can be calculated by referring to Refs. 5 and 6.

The boundary condition (2-19) is replaced by
as $\eta \to \infty$;

$$W_{kV} = W_{kV\infty} \quad (k = 1, 2, \ldots, n - 1) \tag{7.1-6}$$

and the compatibility conditions (2-23), (2-26), and (2-27) at $y = \delta$ are replaced, respectively, by

$$\sum_{k=1}^{n} \dot{m}_{kx} = \dot{m}_x, \tag{7.1-7}$$

$$W_{kV} = W_{kVi} \quad (k = 1, 2, \ldots, n - 1), \tag{7.1-8}$$

$$\rho \sum_{l=1}^{n-1} D_{kl}^+ \left(\frac{\partial W_{lV}}{\partial y} \right)_i = \dot{m}_{kx} - W_{kVi} \dot{m}_x$$
$$(k = 1, 2, \ldots, n - 1). \tag{7.1-9}$$

Equation (2-28) is replaced by

$$W_{kL} = \frac{\dot{m}_{kx}}{\dot{m}_x} \quad (k = 1, 2, \ldots, n) \tag{7.1-10}$$

and the equation of phase equilibrium at the vapor-liquid interface is expressed as

$$X_{kL} = \frac{X_{kVi}\, p}{\gamma_k p_k^\circ(T_i)} \quad (k = 1, 2, \ldots, n), \tag{7.1-11}$$

where X_k denotes the molar fraction of component k, γ_k the activity constant of component k, p the static pressure of the system, and $p_k^\circ(T_i)$ the saturation pressure of component k at T_i.

7.2   Similarity Transformation

The similarity variables are the same as those in Chapter 3, except the following one for the mass conservation of component k :

$$\Phi_{kF}(\eta_{FV}) = \frac{W_{kV} - W_{kV\infty}}{W_{kVi} - W_{kV\infty}} \qquad (7.2\text{-}1)$$

and the similar system of ordinary differential equations together with the boundary and compatibility conditions are derived. The equations peculiar to the multicomponent mixture are written as

$$\Theta''_{FV} + \frac{1}{2}\Pr_V F_{FV}\Theta'_{FV} + \Pr_V \Theta'_{FV}\left(\sum_{k=1}^{n-1}\sum_{l=1}^{n-1} c_{pkn}\frac{W_{lVi} - W_{lV\infty}}{Sc_{kl}}\Phi'_{lF}\right) = 0,$$
$$(7.2\text{-}2)$$

$$\sum_{l=1}^{n-1} a_{kl}\Phi''_{lF} + \frac{1}{2}F_{FV}\Phi'_{kF} = 0 \qquad (k = 1, 2, \ldots, n-1), \qquad (7.2\text{-}3)$$

where

$$Sc_{kl} = \frac{\nu}{D^+_{kl}}, \qquad (7.2\text{-}4)$$

$$a_{kl} = \frac{W_{lVi} - W_{lV\infty}}{(W_{kVi} - W_{kV\infty})Sc_{kl}}, \qquad (7.2\text{-}5)$$

together with the boundary condition

as $\eta \to \infty$:

$$\Phi_{kF\infty} = 0 \qquad (k = 1, 2, \ldots, n-1) \qquad (7.2\text{-}6)$$

and the compatibility conditions

at $\eta_L = \eta_{Li}$:

$$RF_{FLi} = F_{FVi} = 2R\sum_{k=1}^{n}\dot{M}_{kFL} = 2R\dot{M}_{FL}\,; \qquad (7.2\text{-}7)$$

or $\eta_V = 0$:

$$\Phi_{kFi} = 1 \qquad (k = 1, 2, \ldots, n-1), \qquad (7.2\text{-}8)$$

$$\sum_{l=1}^{n-1} a_{kl}\Phi'_{lFi} = c_k \qquad (k = 1, 2, \ldots, n-1), \qquad (7.2\text{-}9)$$

where

$$\dot{M}_{kFL} = \frac{\dot{m}_{kx}x}{\mu_L Re_{Lx}^{\frac{1}{2}}} \qquad (k = 1, 2, \ldots, n), \qquad (7.2\text{-}10)$$

$$\dot{M}_{FL} = \frac{\dot{m}_x x}{\mu_L Re_{Lx}^{\frac{1}{2}}}, \tag{7.2-11}$$

$$c_k = \frac{(\dot{M}_{kFL} - W_{kVi}\dot{M}_{FL})R}{W_{kVi} - W_{kV\infty}} \quad (k = 1, 2, \ldots, n-1). \tag{7.2-12}$$

Equation (7.1-10) is rewritten as

$$W_{kL} = \frac{\dot{M}_{kFL}}{\dot{M}_{FL}} \quad (k = 1, 2, \ldots, n). \tag{7.2-13}$$

7.3   Orthogonal Transformation of the Ordinary Differential Equations using the Matrix Method

The first term in Eq. (7.2-3) contains $(n-2)$ components except component k, i.e., the equations of Φ are composed of $(n-1)$ mutually dependent simultaneous equations. We now transform these equations to $(n-1)$ independent equations.

Equations (7.2-3), boundary condition (7.2-6), and compatibility conditions (7.2-8) and (7.2-9) are expressed by the matrix notation as follows:

$$A\Phi''_F + \frac{1}{2}F_{FV}\Phi'_F = O; \tag{7.3-1}$$

as $\eta_V \to \infty$:

$$\Phi_{F\infty} = O; \tag{7.3-2}$$

at $\eta_V = 0$:

$$\Phi_{Fi} = I, \tag{7.3-3}$$

$$A\Phi'_{Fi} = C; \tag{7.3-4}$$

where

$$\Phi_F = (\Phi_{kF})_1^{n-1}, \tag{7.3-5}$$

$$A = (a_{kl})_{n-1}^{n-1}, \tag{7.3-6}$$

$$C = (c_k)_1^{n-1}, \tag{7.3-7}$$

and O and I are the column matrices of elements 0 and 1, respectively.

The nonsingular matrix A of Eq. (7.3-6) has eigenvalues, which are denoted by $1/Sc_k (k = 1, 2, \ldots, n-1)$, and can be orthogonally transformed to the diagonal matrix, in which the elements consist of the eigenvalues, by using matrix P and its inverse matrix P^{-1} as follows :

$$B = P^{-1}AP, \tag{7.3-8}$$

where

$$\boldsymbol{B} = \left(\frac{\delta_{kl}}{Sc_k} \right)_{n-1}^{n-1}, \tag{7.3-9}$$

$$\boldsymbol{P} = (p_{kl})_{n-1}^{n-1}, \tag{7.3-10}$$

$$\boldsymbol{P}^{-1} = (q_{kl})_{n-1}^{n-1}, \tag{7.3-11}$$

where δ_{kl} is the Kronecker delta, and p_{kl} and q_{kl} can be evaluated by referring to Ref. 7.

$\boldsymbol{\Phi}_F$ in Eqs. (7.3-1)–(7.3-4) is transformed to $\boldsymbol{\Phi}_F^*$ by the equation

$$\boldsymbol{\Phi}_F^* = (\boldsymbol{\Phi}_{kF}^*)_1^{n-1} = \boldsymbol{DP}^{-1}\boldsymbol{\Phi}_F \tag{7.3-12}$$

or

$$\boldsymbol{\Phi}_F = \boldsymbol{PD}^{-1}\boldsymbol{\Phi}_F^*, \tag{7.3-13}$$

where the matrix $\boldsymbol{D}$ is introduced for normalizing the equation corresponding to Eq. (7.3-3) as

$$\boldsymbol{D} = \left(\frac{\delta_{kl}}{\displaystyle\sum_{m=1}^{n-1} q_{km}} \right)_{n-1}^{n-1}. \tag{7.3-14}$$

Substituting Eq. (7.3-13) into Eqs. (7.3-1)–(7.3-4), and using Eq. (7.3-8) $(\boldsymbol{BD}^{-1} = \boldsymbol{D}^{-1}\boldsymbol{B})$, we obtain

$$\boldsymbol{B}\boldsymbol{\Phi}_F^{*\prime\prime} + \frac{1}{2}F_{FV}\boldsymbol{\Phi}_F^{*\prime} = \boldsymbol{O}; \tag{7.3-15}$$

as $\eta_V \rightarrow \infty$:

$$\boldsymbol{\Phi}_{F\infty}^* = \boldsymbol{O}; \tag{7.3-16}$$

at $\eta_V = 0$:

$$\boldsymbol{\Phi}_{Fi}^* = \boldsymbol{I}, \tag{7.3-17}$$

$$\boldsymbol{B}\boldsymbol{\Phi}_{Fi}^{*\prime} = \boldsymbol{DP}^{-1}\boldsymbol{C}. \tag{7.3-18}$$

These equations are expressed with the elements as follows:

$$\Phi_{kF}^{*\prime\prime} + \frac{1}{2}Sc_k F_{FV}\Phi_{kF}^{*\prime} = 0 \qquad (k = 1, 2, \ldots, n-1) ; \tag{7.3-19}$$

as $\eta_V \rightarrow \infty$:

$$\Phi_{kF\infty}^* = 0 \qquad (k = 1, 2, \ldots, n-1) ; \tag{7.3-20}$$

at $\eta_V = 0$:

$$\Phi^*_{kFi} \;=\; 1 \qquad (k = 1, 2, \ldots, n-1)\,, \tag{7.3-21}$$

$$-\Phi^{*\prime}_{kFi} \;=\; \frac{\displaystyle\sum_{m=1}^{n-1} q_{km}\, \frac{W_{mVi}\dot{M}_{FL} - \dot{M}_{mFL}}{W_{mVi} - W_{mV\infty}}}{\displaystyle\sum_{l=1}^{n-1} q_{kl}}\, R\,Sc_k$$
$$(k = 1, 2, \ldots, n-1)\,. \tag{7.3-22}$$

Equation (7.3-13) is expressed as

$$\Phi_{kF} = \sum_{l=1}^{n-1}\sum_{m=1}^{n-1} p_{kl}q_{lm}\,\Phi^*_{lF} \qquad (k = 1, 2, \ldots, n-1)\,. \tag{7.3-23}$$

Equation (7.2-2) is similarly transformed as

$$\Theta''_{FV} + \frac{1}{2}Pr_V F_{FV}\Theta'_{FV} + Pr_V\Theta'_{FV}$$
$$\times \sum_{k=1}^{n-1}\left\{\sum_{l=1}^{n-1} c_{pkn}\,\frac{W_{lVi} - W_{lV\infty}}{Sc_{kl}}\left(\sum_{j=1}^{n-1}\sum_{m=1}^{n-1} p_{lj}q_{jm}\,\Phi^*_{jF}\right)\right\} = 0\,.$$
$$\tag{7.3-24}$$

Equation (7.3-22) can be transformed by using Eqs. (7.2-13) and (7.2-7) as

$$1 - \frac{Sc_k F_{FVi}}{2(\Phi^*_{kFi})} = \frac{1}{W_{kR}} \qquad (k = 1, 2, \ldots, n-1)\,, \tag{7.3-25}$$

where

$$\frac{1}{W_{kR}} = 1 - \frac{\displaystyle\sum_{l=1}^{n-1} q_{kl}}{\displaystyle\sum_{m=1}^{n-1} q_{km}\frac{W_{mVi} - W_{mL}}{W_{mVi} - W_{mV\infty}}} \qquad (k = 1, 2, \ldots, n-1). \tag{7.3-26}$$

7.4 Algebraic Equations for a Multicomponent Vapor

When Φ_{kF} is transformed to Φ^*_{kF} by using Eq. (7.3-12) or Eq. (7.3-23), the system of ordinary differential equations for forced-convection condensation of a multicomponent vapor mixture consists of Eqs. (3.1-10)–(3.1-12), (7.3-24), and (7.3-19) together with the boundary conditions (3.1-15)–(3.1-19) and (7.3-20) and the compatibility conditions (3.1-21), (3.1-22), (7.2-7), (3.1-25)–(3.1-27), (7.3-21), and (7.3-25). This system is identical to that of Eqs. (3.1-10)–(3.1-28) and (3.1-36) in functional form. The solution of the present problem, therefore, can be analogically derived from the solution of the latter, i.e., Φ^*_{kFi}, Θ'_{FLi}, and Θ'_{FVi} are expressed in the functional forms of Eqs. (3.5-6), (3.4-10), and (3.4-14), respectively.

When Φ'_{Fi} and W_R in Eq. (3.5-6) are replaced by $\Phi^{*'}_{kFi}$ in Eqs. (7.3-22) and W_{kR} in Eq. (7.3-26), respectively, we obtain

$$\frac{\sum_{l=1}^{n-1} q_{km} \dfrac{W_{lVi}\dot{M}_{FL} - \dot{M}_{lFL}}{W_{lVi} - W_{lV\infty}}}{\sum_{l=1}^{n-1} q_{kl}} = \frac{C_F(Sc_k)}{RSc_k}\left(\frac{2.5}{1.5 + W_{kR}}\right)^m W_{kR}$$

$$(k = 1, 2, \ldots, n-1), \qquad (7.4\text{-}1)$$

where the exponent m is evaluated by replacing Sc by Sc_k in Eq. (3.5-7).

The value of Eq. (7.3-24) at the vapor-liquid interface is written by using Eqs. (7.3-25) and (7.3-26) as

$$\Theta''_{FVi} + \frac{1}{2}Pr_V F_{FVi}\Theta'_{FVi}(1 - \phi_i) = 0, \qquad (7.4\text{-}2)$$

where

$$\phi_i = \sum_{k=1}^{n-1}\left[\sum_{l=1}^{n-1} c_{pkn}\frac{W_{lVi} - W_{lV\infty}}{Sc_{kl}}\left\{\sum_{j=1}^{n} p_{lj}Sc_j\left(\sum_{m=1}^{n-1} q_{jm}\frac{W_{mVi} - W_{mL}}{W_{mVi} - W_{mV\infty}}\right)\right\}\right]. \qquad (7.4\text{-}3)$$

By referring to the modification from Eq. (3.4-14) to Eq. (3.4-16), we obtain the equation

$$\frac{-\Theta'_{FVi}}{C_F(Pr_V)} = 1 + 2.6Pr_V^{0.66}\dot{M}_{FV}^{1.05}\left(1 - \frac{2}{3}\phi_i\right). \qquad (7.4\text{-}4)$$

Substituting Eqs. (3.4-10), (7.4-4), and (7.2-7) into Eq. (3.1-27), we obtain the equation

$$0.433\left(1.367 - \frac{0.432}{\sqrt{2R\dot{M}_{FL}}} + \frac{1}{2R\dot{M}_{FL}}\right)^{\frac{1}{2}}(1 + 0.320H_i^{0.87})^{-1}$$

$$= \frac{Pr_L\dot{M}_{FL}}{H_i} + \frac{\lambda_V}{\lambda_L}\left(\frac{\nu_L}{\nu_V}\right)^{\frac{1}{2}}\frac{(T_{V\infty} - T_i)}{(T_i - T_w)}$$

$$\times C_F(Pr_V)\left\{1 + 2.6Pr_V^{0.66}(R\dot{M}_{FL})^{1.05}\left(1 - \frac{2}{3}\phi_i\right)\right\}. \qquad (7.4\text{-}5)$$

By simultaneously solving Eqs. (7.4-1), (7.4-5), (7.2-13), and (7.1-11) for a given substance at given wall temperature and bulk conditions, we can obtain the values of T_i, $\dot{M}_{FL}$, $\dot{M}_{kFL}$, W_{kL} and $W_{kVi}(k = 1, 2, ..., n - 1)$, and compute the values of $\dot{m}_{kx}$, $\dot{m}_x$, q_{wx}, q_{cx}, etc., from the respective definition equations.

7.5   Algebraic Equations for a Ternary Vapor

We reduce the algebraic equations for a ternary vapor mixture from those for a multicomponent vapor mixture in the previous section in preparation for the computation in the next section.

The value of D_{kl}^M in Eq. (7.1-5) is given for $k \neq \ell$ and $m \neq k, \ell$ by :

$$D_{kl}^M = D_{kl}^B \left\{ 1 + \frac{X_m \left(\dfrac{M_m D_{km}^B}{M_l} - D_{kl}^B \right)}{X_k D_{\ell m}^B + X_l D_{mk}^B + X_m D_{kl}^B} \right\} ; \qquad (7.5\text{-}1a)$$

and for $k = \ell$ by :

$$D_{kl}^M = 0 , \qquad (7.5\text{-}1b)$$

where D_{kl}^B is the diffusivity between components k and ℓ in a binary vapor mixture and X_k is the molar fraction of component k.

Matrices $\boldsymbol{A}$, $\boldsymbol{B}$, $\boldsymbol{P}$, $\boldsymbol{P}^{-1}$, $\boldsymbol{C}$, and $\boldsymbol{D}$ are reduced to

$$\boldsymbol{A} = (a_{kl}) = \begin{pmatrix} \dfrac{1}{Sc_{11}} & \dfrac{W_{2Vi} - W_{2V\infty}}{W_{1Vi} - W_{1V\infty}} Sc_{12} \\[4mm] \dfrac{W_{1Vi} - W_{1V\infty}}{W_{2Vi} - W_{2V\infty}} Sc_{21} & \dfrac{1}{Sc_{22}} \end{pmatrix} , \quad (7.5\text{-}2)$$

$$\boldsymbol{B} = \begin{pmatrix} \dfrac{1}{Sc_1} & 0 \\[3mm] 0 & \dfrac{1}{Sc_2} \end{pmatrix} , \qquad (7.5\text{-}3)$$

where the eigenvalues $1/Sc_k$ of matrix $\boldsymbol{A}$ are the roots of the following quadratic equation (either root will do for $k = 1$) :

$$Sc_{11} Sc_{22}\, t^2 - (Sc_{11} + Sc_{22})t + \left(1 - \frac{Sc_{11} Sc_{22}}{Sc_{12} Sc_{21}} \right) = 0 , \qquad (7.5\text{-}4)$$

$$\boldsymbol{P} = (p_{kl}) = \begin{pmatrix} \dfrac{1}{\sqrt{1+\epsilon^2}} & \dfrac{\zeta}{\sqrt{1+\zeta^2}} \\[3mm] \dfrac{\epsilon}{\sqrt{1+\epsilon^2}} & \dfrac{1}{\sqrt{1+\zeta^2}} \end{pmatrix} , \tag{7.5-5}$$

$$\boldsymbol{P}^{-1} = (q_{kl}) = \dfrac{1}{1-\epsilon\zeta} \begin{pmatrix} \sqrt{1+\epsilon^2} & -\zeta\sqrt{1+\zeta^2} \\[3mm] -\epsilon\sqrt{1+\epsilon^2} & \sqrt{1+\zeta^2} \end{pmatrix} , \tag{7.5-6}$$

where

$$\epsilon = -\dfrac{Sc_1 Sc_{22}(W_{1Vi} - W_{1V\infty})}{Sc_{21}(Sc_1 - Sc_{22})(W_{2Vi} - W_{2V\infty})} , \tag{7.5-7}$$

$$\zeta = -\dfrac{Sc_2 Sc_{11}(W_{2Vi} - W_{2V\infty})}{Sc_{12}(Sc_2 - Sc_{11})(W_{1Vi} - W_{1V\infty})} , \tag{7.5-8}$$

$$\boldsymbol{C} = (c_k) = \begin{pmatrix} \dfrac{(1 - W_{1Vi})\dot{M}_{1FL} - W_{1Vi}(\dot{M}_{2FL} - \dot{M}_{3FL})}{W_{1Vi} - W_{1V\infty}} R \\[4mm] \dfrac{(1 - W_{2Vi})\dot{M}_{2FL} - W_{2Vi}(\dot{M}_{3FL} - \dot{M}_{1FL})}{W_{2Vi} - W_{2V\infty}} R \end{pmatrix} , \tag{7.5-9}$$

$$\boldsymbol{D} = \begin{pmatrix} \dfrac{1}{q_{11} + q_{12}} & 0 \\[4mm] 0 & \dfrac{1}{q_{21} + q_{22}} \end{pmatrix} . \tag{7.5-10}$$

Equations (7.4-1) and (7.4-5) are reduced to

$$\dfrac{q_{11}\dfrac{W_{1Vi}\dot{M}_{FL} - \dot{M}_{1FL}}{W_{1Vi} - W_{1V\infty}} + q_{12}\dfrac{W_{2Vi}\dot{M}_{FL} - \dot{M}_{2FL}}{W_{2Vi} - W_{2V\infty}}}{(q_{11} + q_{12})}$$
$$= \dfrac{C_F(Sc_1)}{RSc_1}\left(\dfrac{2.5}{1.5 + W_{1R}}\right)^m W_{1R} , \tag{7.5-11a}$$

$$\dfrac{q_{21}\dfrac{W_{1Vi}\dot{M}_{FL} - \dot{M}_{1FL}}{W_{1Vi} - W_{1V\infty}} + q_{22}\dfrac{W_{2Vi}\dot{M}_{FL} - \dot{M}_{2FL}}{W_{2Vi} - W_{2V\infty}}}{(q_{21} + q_{22})}$$
$$= \dfrac{C_F(Sc_2)}{RSc_2}\left(\dfrac{2.5}{1.5 + W_{2R}}\right)^m W_{2R} , \tag{7.5-11b}$$

$$0.433 \left(1.367 - \frac{0.432}{\sqrt{2R\dot{M}_{FL}}} + \frac{1}{2R\dot{M}_{FL}} \right)^{\frac{1}{2}} \left(1 + 0.320 H_i^{0.87} \right)^{-1}$$

$$= \frac{\mathrm{Pr}_L \dot{M}_{FL}}{H_i} + \frac{\lambda_V}{\lambda_L} \left(\frac{\nu_L}{\nu_V} \right)^{\frac{1}{2}} \frac{(T_{V\infty} - T_i)}{(T_i - T_w)}$$

$$\times C_F(\mathrm{Pr}_V) \left\{ 1 + 2.6\mathrm{Pr}_V^{0.66} \left(R\dot{M}_{FL} \right)^{1.05} \left(1 - \frac{2}{3}\phi_i \right) \right\}, \quad (7.5\text{-}12)$$

where

$$\frac{1}{W_{1R}} = 1 - \frac{(q_{11} + q_{12})}{q_{11} \dfrac{W_{1Vi} - W_{1L}}{W_{1Vi} - W_{1V\infty}} + q_{12} \dfrac{W_{2Vi} - W_{1L}}{W_{2Vi} - W_{2V\infty}}}, \quad (7.5\text{-}13a)$$

$$\frac{1}{W_{2R}} = 1 - \frac{(q_{21} + q_{22})}{q_{21} \dfrac{W_{1Vi} - W_{1L}}{W_{1Vi} - W_{1V\infty}} + q_{22} \dfrac{W_{2Vi} - W_{1L}}{W_{2Vi} - W_{2V\infty}}}. \quad (7.5\text{-}13b)$$

The ten unknown values of W_{1Vi}, W_{2Vi}, W_{1L}, W_{2L}, W_{1R}, W_{2R}, $\dot{M}_{1FL}$, $\dot{M}_{2FL}$, $\dot{M}_{FL}$, and T_i can be solved from ten simultaneous equations composed of the above five equations: three of Eq. (7.1-11) and two of Eq. (7.2-13).

7.6   An Example for an Air-Methanol-Water Mixture[3]

We take an air-methanol-water mixture as an example, because the physical properties of each component are well known, and the computation is made for $p = 101.325$ kPa, $T_{V\infty} = 363.15$ K, and variable bulk concentration under the assumptions that the convection heat transfer in the vapor boundary layer and the convection term for the condensate film are negligible, the bulk vapor is saturated, and the value of the exponent m is constant for large values of R. The physical properties used are shown in the Appendix, and the representative values are taken according to the conclusion in Chapter 9.

Figure 7.6-1 shows the relevant phase equilibrium diagram at $p = 101.325$ kPa, where the subscripts 1, 2, and 3 denote air, methanol, and water, respectively. For example, when $T_{V\infty} = 90°C$ and $W_{1V\infty} = 0.2$ in the bulk and $T_i = 74.8°C$ and $W_{1Vi} = 0.4$ at the vapor-liquid interface, the value of $W_{2V\infty}/(1 - W_{1V\infty})$, $W_{2Vi}/(1 - W_{1Vi})$, and W_{2L} are represented by points

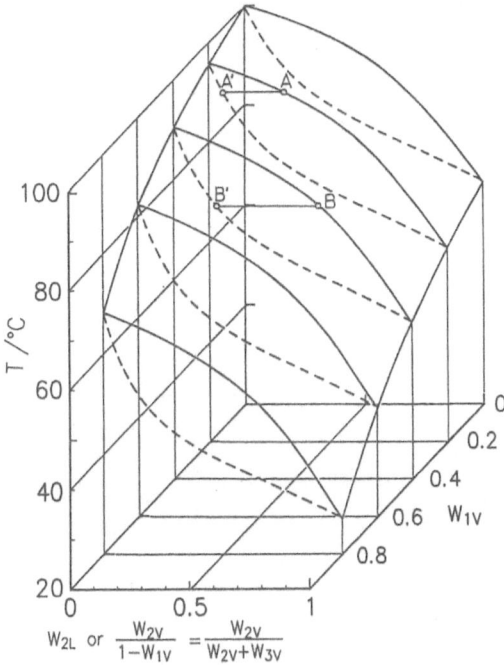

FIGURE 7.6-1. Phase equilibrium diagram for an air-methanol-water mixture at $p=101.325$ kPa and $T_\infty=363.15$ K. Subscripts 1: air; 2: methanol; 3: water.

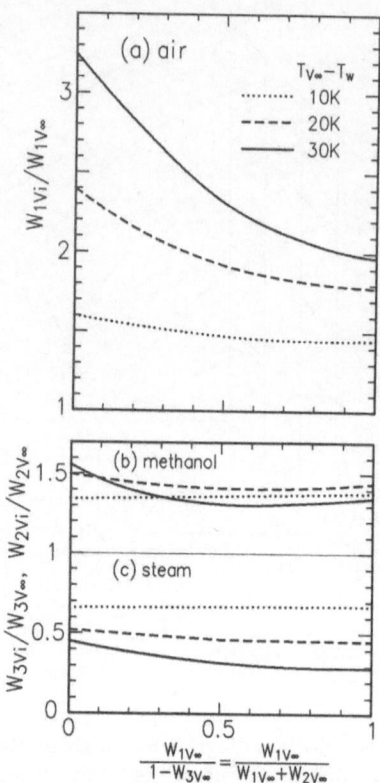

FIGURE 7.6-2.   Relation of $W_{kVi}/W_{kV\infty}$ versus $W_{1V\infty}/(1 - W_{3V\infty})$ for an air-methanol-water mixture at $p = 101.325$ kPa and $T_\infty = 363.15$ K.

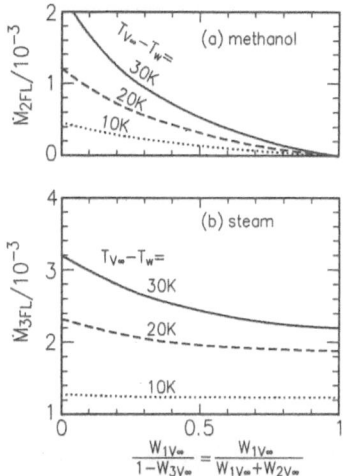

FIGURE 7.6-3. Relation of $\dot{M}_{2FL}$ and $\dot{M}_{3FL}$ versus $W_{1V\infty}/(1 - W_{3V\infty})$ for an air-methanol-water mixture at $p = 101.325$ kPa and $T_{V\infty} = 363.15$ K.

A, B, and B', respectively, and the state of the vapor mixture varies from point A to B in the vapor boundary layer.

Figures 7.6-2(a), (b), and (c) show the relation of $W_{kVi}/W_{kV\infty}$ ($k = 1, 2$, and 3) versus $W_{1V\infty}/(1 - W_{3V\infty})$ with $T_{V\infty} - T_w$ as a parameter, where (a), (b), and (c) correspond to the concentration ratios of air, methanol vapor, and steam, respectively. The abscissa in each figure is so taken that the values of 0 and 1 correspond to the binary vapor mixtures of methanol-water and air-water, respectively, because any air-methanol mixture without water does not exist under the prescribed bulk condition. We can see that noncondensable air and volatile methanol vapor are concentrated ($W_{1Vi}/W_{1V\infty}>1$, $W_{2Vi}/W_{2V\infty}>1$) and less volatile steam is diluted ($W_{3Vi}/W_{3V\infty}<1$) at the vapor-liquid interface. Generally, the values of $W_{kVi}/W_{kV\infty}$ increase and the variation with $W_{1V\infty}/(1 - W_{3V\infty})$ becomes marked as $T_{V\infty} - T_w$ increases.

Figure 7.6-3 (a) and (b) show the relations of $\dot{M}_{2FL}$ and $\dot{M}_{3FL}$ versus $W_{1V\infty}/(1 - W_{3V\infty})$, respectively. The values of $\dot{M}_{2FL}$ and $\dot{M}_{3FL}$ and their decreasing rate with $W_{1V\infty}/(1 - W_{3V\infty})$ are large as the values of $T_{V\infty} - T_w$ increase. We can see that the concentration of methanol in the condensate increases for a small value of $W_{1V\infty}/(1 - W_{3V\infty})$ and a large value of $T_{V\infty} - T_w$.

Figure 7.6-4 shows the relation of $(T_i - T_w)/(T_{V\infty} - T_w)$ versus $W_{1V\infty}/(1 - W_{3V\infty})$. The values of $(T_i - T_w)/(T_{V\infty} - T_w)$ become large as $T_{V\infty} - T_w$ increases, and decrease with the increase of $W_{1V\infty}/(1 - W_{3V\infty})$. These results correspond to the variation of thermal resistance across the condensate film according to the variation of condensation mass flux.

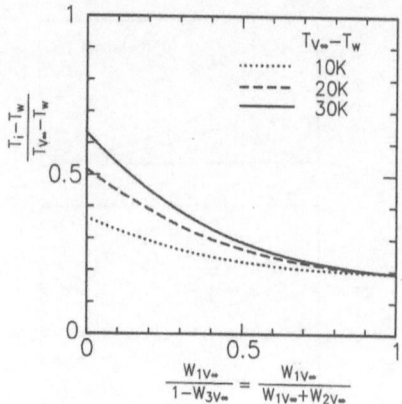

FIGURE 7.6-4. Relation of $(T_i - T_w)/(T_{V\infty} - T_w)$ versus $W_{1V\infty}/(1 - W_{3V\infty})$ for an air-methanol-water mixture at $p = 101.325$ kPa and $T_{V\infty} = 363.15$ K.

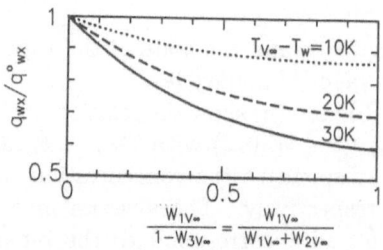

FIGURE 7.6-5. Effect of air upon condensation heat flux. The heat flux q_{wx} is for air-methanol-water and q_{wx}^0 is for methanol-water in the case at p=101.325 kPa and $T_{V\infty}$=363.15 K.

Figure 7.6-5 shows the effect of air upon the heat flux at the cooling surface, where q_{wx} and q_{wx}^0 denote the heat fluxes for the mixtures of air-methanol-water and methanol-water, respectively. The values of q_{wx}/q_{wx}^0 decrease with the increase of $W_{1V\infty}/(1 - W_{3V\infty})$ and the decreasing rate is large as $T_{V\infty} - T_w$ is large, which corresponds to the variation of total condensation mass flux $\dot{M}_{FL}$. Incidentally, in this case, the effect of the aforementioned neglected terms upon the heat flux is only a few percent .

REFERENCES

1. Toor, H. L., Solution of the Linearized Equations of Multicomponent Mass Transfer: II. Matrix Methods, *AIChE J.*, 10-4, 460–465 (1964).

2. Stewart, W. E. and R. Prober, Matrix Calculation of Multicomponent Mass Transfer in Isothermal Systems, *I & EC Fundament.*, 3-3, 224–235 (1964).

3. Fujii, T. and Sh. Koyama, Laminar Forced Convection Condensation of a Ternary Vapor Mixture on a Flat Plate, *Fundamentals of Phase Change: Boiling and Condensation—ASME HTD*, 38, 81–87 (1984).

4. Koyama, Sh., M. Goto, and T. Fujii, Laminar Film Condensation of Multicomponent Mixtures on a Flat Plate—First Report, Forced Convection Condensation (in Japanese), *Reports of Institute of Advanced Material Study, Kyushu University*, 1–1, 77–83 (1987).

5. Bird, R. B., W. E. Stewart, and E. N. Lightfoot, "Transport Phenomena," p.570, John Wiley & Sons, New York (1960).

6. Hirshfelder, J. O., C. F. Curtiss, and R. B. Bird, "Molecular Theory of Gases and Liquids," John Wiley & Sons, New York (1954)

7. Wylie, C. R. Jr., "Advanced Engineering Mathematics," pp.429–492, 3rd Ed., McGraw-Hill, New York (1966).

8

Free-Convection Condensation of Multicomponent Vapors

From the results for the similarity transformation in Chapter 4 and for the orthogonal transformation in Chapter 7 we can analogically infer the algebraic equations for the free-convection condensation of a multicomponent vapor mixture. In the inference, however, we have no sound basis for treating the buoyancy term. Therefore, we compare the formulated results with those derived from the similarity solutions for free-convection heat and mass transfer and free-convection condensation of a ternary vapor mixture.

In the present case the buoyancy term in Eq. (2-5) is expressed as

$$1 - \frac{\rho_\infty}{\rho} = 1 - \{1 - \Omega_T (T_{V\infty} - T)\} \left\{ 1 - \sum_{k=1}^{n-1} \Omega_{Wk} (W_{kV} - W_{kV\infty}) \right\}$$
(8-1a)

$$\approx \Omega_T (T_{V\infty} - T) + \sum_{k=1}^{n-1} \Omega_{Wk} (W_{kV} - W_{kV\infty}) ,$$
(8-1b)

where

$$\Omega_T = \frac{1}{T_{V\infty}} ,$$
(8-2)

$$\Omega_{Wk} = \frac{\left(\dfrac{1}{M_n} - \dfrac{1}{M_k} \right)}{\left\{ \displaystyle\sum_{l=1}^{n} \dfrac{W_{lV\infty}}{M_l} \right\}} \qquad (k = 1, 2, \ldots, n-1) .$$
(8-3)

The normalized concentration Φ_k is defined by Eq. (7.2-1), where the subscript F is changed to G.

8.1 Basic Equations, Transformations, and Algebraic Equations[1]

The basic equations together with the boundary and compatibility conditions, which are derived by the similarity and orthogonal transformations, are the same as those obtained in Chapters 4 and 7. The exception is the buoyancy term in the equation of momentum conservation and the boundary and compatibility conditions for W_{kV}, i.e., the equations of momentum and energy of condensate are expressed by Eqs. (4.1-5) and (4.1-6), respectively. The equations of momentum, energy, and mass of component k for the vapor boundary layer are expressed as

$$F_{GV}''' + 3F_{GV}F_{GV}'' - 2F_{GV}'^2 + \omega_T\Theta_{GV} + \sum_{k=0}^{n-1}\omega_k^*\Phi_{kG}^* = 0, \qquad (8.1\text{-}1)$$

$$\Theta_{GV}'' + 3Pr_V F_{GV}\Theta_{GV}' = 0, \qquad (8.1\text{-}2)$$

$$\Phi_{kG}^{*''} + 3Sc_k F_{GV}\Phi_{kG}^{*'} = 0$$
$$(k = 1, 2, \ldots, n-1), \qquad (8.1\text{-}3)$$

where

$$\omega_T = \Omega_T(T_{V\infty} - T_i), \qquad (8.1\text{-}4)$$

$$\omega_k^* = \sum_{l=1}^{n-1}\sum_{m=1}^{n-1}\Omega_{Wl}(W_{lVi} - W_{lV\infty})p_{lk}q_{km}$$
$$(k = 1, 2, \ldots, n-1), \qquad (8.1\text{-}5)$$

the terms of the product of ω_T and ω_k^* in Eq. (8.1-1) are neglected, and the diffusion term in Eq. (8.1-2) is also neglected.

As for the boundary conditions, Eqs. (4.1-12)–(4.1-16) are valid and as $\eta_V \to \infty$:

$$\Phi_{kG\infty}^* = 0 \quad (k = 1, 2, \ldots, n-1). \qquad (8.1\text{-}6)$$

As for compatibility conditions, Eqs. (4.1-18), (4.1-19), and (4.1-22)–(4.1-24) are valid and at $\eta_V = 0$:

$$F_{GVi} = RF_{GLi} = \frac{R}{3}\sum_{k=1}^{n}\dot{M}_{kGL} = \frac{R\dot{M}_{GL}}{3}, \qquad (8.1\text{-}7)$$

$$\Phi_{kGi}^* = 1 \quad (k = 1, 2, \ldots, n-1), \qquad (8.1\text{-}8)$$

$$\Phi_{kGi}^* = RSc_k\frac{\displaystyle\sum_{m=1}^{n-1}q_{km}\frac{\dot{M}_{mGL} - W_{mVi}\dot{M}_{GL}}{W_{mVi} - W_{mV\infty}}}{\displaystyle\sum_{l=1}^{n-1}q_{kl}}$$
$$(k = 1, 2, \ldots, n-1), \qquad (8.1\text{-}9)$$

and $\dot{M}_{kGL}$, $\dot{M}_{GL}$, and W_{kL} are expressed by Eqs. (4.1-27), (4.1-28), and (4.1-30), respectively, where k is taken from 1 to $n-1$.

Equation (8.1-9) is transformed by using Eqs. (4.1-30) and (8.1-7) as

$$1 - \frac{3Sc_k F_{GVi}}{-\Phi^{*'}_{ki}} = \frac{1}{W_{kR}} \qquad (k = 1, 2, \ldots, n-1).\qquad (8.1\text{-}10)$$

This is the same form as Eq. (4.1-31) and W_{kR} is given by Eq. (7.3-26).

Since the equations together with the above-mentioned boundary and compatibility conditions are the same in functional form as in Chapter 4, the solutions of the relevant problem should be expressed as follows:

$$\frac{-\Phi^{*'}_{kGi}}{\sqrt{2}C_G(Sc_k)\,(\chi_{ki}Sc_k)^{\frac{1}{4}}} = \left(\frac{2}{1+W_{kR}}\right)^{0.5} W_{kR}^n$$
$$(k = 1, 2, \ldots, n-1),\qquad (8.1\text{-}11)$$

$$\frac{-\Theta^{*'}_{GVi}}{\sqrt{2}C_G(Pr_V)\,(\chi_i^+ Pr_V)^{\frac{1}{4}}} = 1 + kPr_V^{0.66}\left(\omega_i^{-\frac{1}{4}} R\dot{M}_{GL}\right)^l,\qquad (8.1\text{-}12)$$

where

$$\chi_{ki} = \omega_T\left(\frac{Sc_k}{Pr_V}\right)^{\frac{1}{2}} + \sum_{l=1}^{n-1}\omega_l^*\left(\frac{Sc_k}{Sc_l}\right)^{\frac{1}{2}} \quad (k = 1, 2, \ldots, n-1),\qquad (8.1\text{-}13)$$

$$\chi_i^+ = \omega_T + \sum_{l=1}^{n-1}\omega_l^*\left(\frac{Pr_V}{Sc_l}\right)^{\frac{1}{2}},\qquad (8.1\text{-}14)$$

$$\omega_i = \omega_T + \sum_{l=1}^{n-1}\omega_l^*.\qquad (8.1\text{-}15)$$

As for Θ'_{GLw} and Θ'_{GLi}, Eq. (4.4-6) is valid. Equations (8.1-13), (8.1-14), and (8.1-15) have been analogically proposed by referring to the results for binary and ternary vapor mixtures, the latter of which is explained later.

Substituting Eqs. (8.1-7) and (8.1-11) into Eq. (8.1-10) yields

$$\frac{\sqrt{2}C_G(Sc_k)\,(\chi_{ki}Sc_k)^{\frac{1}{4}}}{Sc_k R}\left(\frac{2}{1+W_{kR}}\right)^{0.5} W_{kR}^n$$
$$= \frac{\displaystyle\sum_{m=1}^{n-1} q_{km}\frac{W_{mVi}\dot{M}_{GL} - \dot{M}_{kGL}}{W_{mVi} - W_{mV\infty}}}{\displaystyle\sum_{l=1}^{n-1} q_{kl}} \qquad (k = 1, 2, \ldots, n-1)\quad (8.1\text{-}16)$$

and substituting Eqs. (8.1-12) and (4.4-4) into Eq. (4.1-24) yields

$$\dot{M}_{GL}^{-\frac{1}{3}} = \frac{Pr_L \dot{M}_{GL}}{Hi} + \frac{\lambda_V}{\lambda_L}\left(\frac{\nu_L}{\nu_V}\right)^{\frac{1}{2}}\left(\frac{T_{V\infty} - T_i}{T_i - T_w}\right)\sqrt{2}C_G(Pr_V)$$
$$\times (\chi_i^+ Pr_V)^{\frac{1}{4}}\left\{1 + kPr_V^{0.66}\left(\omega_i^{-\frac{1}{4}}R\dot{M}_{GL}\right)^l\right\}. \qquad (8.1\text{-}17)$$

From Eqs. (8.1-16), (8.1-17), (7.2-13), and (7.1-11) we can determine the relevant physical quantities.

8.2 Free-Convection Heat and Mass Transfer of Ternary Vapors

This case corresponds to a limiting case where the condensate film thickness becomes negligibly small. Accordingly, the equations in the vapor boundary layer together with the boundary and compatibility conditions in Section 8.1 is valid when the subscript i is replaced by s, except that

at $\eta = 0$:
$$F_s' = 0, \qquad (8.2\text{-}1)$$
$$F_s = 0. \qquad (8.2\text{-}2)$$

Therefore, the system of the ordinary differential equations for the relevant problem is written as

$$F''' + 3F''F - 2F'^2 + \omega_T\Theta + \omega_1^*\Phi_1^* + \omega_2^*\Phi_2^* = 0, \qquad (8.2\text{-}3)$$
$$\Theta'' + 3Pr_V F\Theta' = 0, \qquad (8.2\text{-}4)$$
$$\Phi_k^{*''} + 3Sc_k F\Phi_k^{*'} = 0 \quad (k = 1, 2); \quad (8.2\text{-}5)$$

at $\eta = 0$:
$$\Phi_{ks}^* = 1 \quad (k = 1, 2); \qquad (8.2\text{-}6)$$

as $\eta \to \infty$:
$$\Phi_{k\infty}^* = 0 \quad (k = 1, 2); \qquad (8.2\text{-}7)$$

where

$$\omega_1^* = \{\Omega_{W1}(W_{1s} - W_{1\infty}) + \Omega_{W2}(W_{2s} - W_{2\infty})\epsilon\}\frac{(1-\zeta)}{(1-\epsilon\zeta)}, \qquad (8.2\text{-}8)$$

$$\omega_2^* = \{\Omega_{W1}(W_{1s} - W_{1\infty})\zeta + \Omega_{W2}(W_{2s} - W_{2\infty})\}\frac{(1-\epsilon)}{(1-\epsilon\zeta)}, \qquad (8.2\text{-}9)$$

$$\Phi_k^* = \frac{W_k^* - W_{k\infty}^*}{W_{ks}^* - W_{k\infty}^*} \quad (k = 1, 2), \qquad (8.2\text{-}10)$$

$$W_k^* = \frac{q_{k1}W_1 + q_{k2}W_2}{q_{k1} + q_{k2}} \quad (k = 1, 2). \qquad (8.2\text{-}11)$$

To reduce the number of parameters in Eq. (8.2-3) we introduce the following variables:

$$\tilde{\eta} = \left(\frac{g\omega_s}{4\nu_V^2 x}\right)^{\frac{1}{4}} y = \omega_s^{\frac{1}{4}}\eta, \qquad (8.2\text{-}12)$$

$$\Psi = 2\sqrt{2}\left(g\omega_s\nu_s^2 x^3\right)^{\frac{1}{4}}\tilde{F}(\tilde{\eta}), \qquad (8.2\text{-}13)$$

where

$$\omega_s = \omega_T + \Omega_{W1}(W_{1s} - W_{1\infty}) + \Omega_{W2}(W_{2s} - W_{2\infty}) \quad (8.2\text{-}14a)$$
$$= \omega_T + \omega_1 + \omega_2 = \omega_T + \omega_1^* + \omega_2^*. \qquad (8.2\text{-}14b)$$

Then, Eqs. (8.2-3), (8.2-4), and (8.2-5) are, respectively, changed to

$$\tilde{F}''' + 3\tilde{F}''\tilde{F} - 2\tilde{F}'^2 + \frac{\Theta + N_1\tilde{\Phi}_1^* + N_2\tilde{\Phi}_2^*}{1 + N_1 + N_2} = 0, \qquad (8.2\text{-}15)$$

$$\tilde{\Theta}'' + 3Pr_V\tilde{F}\tilde{\Theta}' = 0, \qquad (8.2\text{-}16)$$

$$\tilde{\Phi}_k^{*''} + 3Sc_k\tilde{F}\tilde{\Phi}^{*'} = 0 \quad (k = 1, 2), \qquad (8.2\text{-}17)$$

where

$$N_k = \frac{\omega_k^*}{\omega_T} \qquad (k = 1, 2). \qquad (8.2\text{-}18)$$

Watabe et al.[2] solved Eqs. (8.2-15), (8.2-16), and (8.2-17) subject to conditions (8.2-1), (8.2-2), (4.1-15), (8.2-6), (8.2-7), (4,1-23), and (4.1-16) for the given parameters of Pr_V, N_1, N_2, Sc_1, and Sc_2. Some examples of the results are shown in Table 8.2-1.

The diffusion mass flux j_k^*, local mass transfer coefficient β_{kx}^*, and Sherwood number Sh_{kx}^* are defined as

$$j_k^* = -\rho_V D_k^*\left(\frac{\partial W_k^*}{\partial y}\right)_s = \beta_{kx}^*\left(W_{ks}^* - W_{k\infty}^*\right)$$
$$(k = 1, 2), \quad (8.2\text{-}19)$$

$$Sh_{kx}^* = \frac{\beta_{kx}^* x}{\rho_V D_k^*} \qquad (k = 1, 2), \qquad (8.2\text{-}20)$$

where

$$D_k^* = \frac{\nu_V}{Sc_k} \qquad (k = 1, 2). \qquad (8.2\text{-}21)$$

Then, Sh_{kx}^* is expressed as

TABLE 8.2-1. Boundary values of the similarity solution and the values calculated from Eqs. (8.2-26) and (8.2-35) for the single phase free-convection heat and mass transfer of a ternary mixture.

No.	Pr_V	N_1	N_2	Sc_1	Sc_2	F''_s	$-\Phi^{*\prime}_{1s}$	$-\Phi^{*\prime}_{2s}$	$-\Theta'_s$	$-\Phi^{*\prime}_{1s}$	$-\Phi^{*\prime}_{2s}$	$-\Theta'_s$
							similarity solutions			cal. from Eq.(8.2-26,35)		
1-1				0.6	0.6	0.6946	0.4721	0.4721	0.4721	0.4718	0.4718	0.4718
1-2				0.6	1.4	0.6737	0.4603	0.6907	0.4603	0.4613	0.6909	0.4613
1-3	0.6	0.5	0.5	1.0	0.6	0.6818	0.5971	0.4644	0.4644	0.5950	0.4650	0.4650
1-4				1.0	1.4	0.6606	0.5830	0.6809	0.4515	0.5809	0.6800	0.4540
1-5				1.4	0.6	0.6737	0.6907	0.4603	0.4603	0.6909	0.4613	0.4613
1-6				1.4	1.4	0.6524	0.6752	0.6752	0.4470	0.6739	0.6739	0.4500
2-1				0.6	0.6	0.6946	0.4721	0.4721	0.4721	0.4718	0.4718	0.4718
2-2				0.6	1.4	0.6807	0.4644	0.6956	0.4644	0.4649	0.6962	0.4649
2-3	0.6	1.5	0.5	1.0	0.6	0.6689	0.5881	0.4559	0.4559	0.5859	0.4579	0.4579
2-4				1.0	1.4	0.6546	0.5781	0.6759	0.4467	0.5762	0.6744	0.4503
2-5				1.4	0.6	0.6524	0.6752	0.4470	0.4470	0.6739	0.4500	0.4500
2-6				1.4	1.4	0.6378	0.6638	0.6638	0.4368	0.6619	0.6619	0.4419
3-1				0.6	0.6	0.6946	0.4721	0.4721	0.4721	0.4718	0.4718	0.4718
3-2				0.6	1.4	0.6631	0.4539	0.6832	0.4539	0.4557	0.6825	0.4557
3-3	0.6	1.5	1.5	1.0	0.6	0.6754	0.5927	0.4603	0.4603	0.5905	0.4615	0.4615
3-4				1.0	1.4	0.6431	0.5698	0.6672	0.4392	0.5684	0.6653	0.4442
3-5				1.4	0.6	0.6631	0.6832	0.4539	0.4539	0.6825	0.4557	0.4557
3-6				1.4	1.4	0.6304	0.6576	0.6576	0.4312	0.6556	0.6556	0.4377
4-1				0.6	0.6	0.6946	0.4721	0.4721	0.4721	0.4718	0.4718	0.4718
4-2				0.6	1.4	0.7356	0.4925	0.7301	0.4925	0.4910	0.7353	0.4910
4-3	0.6	0.5	-0.5	1.0	0.6	0.6689	0.5881	0.4559	0.4559	0.5859	0.4579	0.4579
4-4				1.0	1.4	0.7107	0.6142	0.7144	0.4794	0.6126	0.7170	0.4787
4-5				1.4	0.6	0.6524	0.6752	0.4470	0.4470	0.6739	0.4500	0.4500
4-6				1.4	1.4	0.6946	0.7049	0.7049	0.4720	0.7066	0.7066	0.4718
5-2				0.6	0.6	0.6946	0.4721	0.4721	0.4721	0.4718	0.4718	0.4718
5-3				0.6	1.4	0.7152	0.4827	0.7179	0.4827	0.4817	0.7214	0.4817
5-4	0.6	1.5	-0.5	1.0	0.6	0.6557	0.5782	0.4465	0.4465	0.5764	0.4505	0.4505
5-5				1.0	1.4	0.6771	0.5930	0.6919	0.4602	0.5908	0.6915	0.4617
5-6				1.4	0.6	0.6304	0.6576	0.4312	0.4312	0.6556	0.4377	0.4377
5-7				1.4	1.4	0.6524	0.6752	0.6752	0.4470	0.6739	0.6739	0.4500

$$Sh^*_{kx} = \frac{1}{\sqrt{2}} \left(-\tilde{\Phi}^{*'}_{ks} \right) Gr^{\frac{1}{4}}_x$$

$$= \frac{1}{\sqrt{2}} \left(-\Phi^{*'}_{ks} \right) \left(\frac{gx^3}{\nu_V{}^2} \right)^{\frac{1}{4}} \qquad (k = 1, 2), \qquad (8.2\text{-}22)$$

where

$$Gr_x = \frac{g\omega_s x^3}{\nu_V{}^2}. \qquad (8.2\text{-}23)$$

Considering the functional form of Eqs. (4.5-2) and (4.5-3), we assume the following correlation equation for Sh^*_{kx} :

$$Sh^*_{kx} = C_G(Sc_k) \left\{ \left(\frac{g\chi_{ki} x^3}{\nu_V{}^2} \right) Sc_k \right\}^{\frac{1}{4}} \qquad (k = 1, 2), \qquad (8.2\text{-}24)$$

where

$$\chi_{ki} = \omega_T \left(\frac{Sc_k}{Pr_V} \right)^{\frac{1}{2}} + \omega^*_1 \left(\frac{Sc_k}{Sc_1} \right)^{\frac{1}{2}} + \omega^*_2 \left(\frac{Sc_k}{Sc_2} \right)^{\frac{1}{2}} \qquad (8.2\text{-}25a)$$

$$= \frac{\left(\frac{Sc_k}{Pr_V} \right)^{\frac{1}{2}} + N_1 \left(\frac{Sc_k}{Sc_1} \right)^{\frac{1}{2}} + N_2 \left(\frac{Sc_k}{Sc_2} \right)^{\frac{1}{2}}}{1 + N_1 + N_2} \omega_s \quad (k = 1, 2), \quad (8.2\text{-}25b)$$

The comparison between Eqs. (8.2-22) and (8.2-24) gives the equation

$$-\Phi^{*'}_{ks} = \sqrt{2} C_G(Sc_k)(\chi_{ki} Sc_k)^{\frac{1}{4}} = \omega_s{}^{\frac{1}{4}} \left(-\tilde{\Phi}^{*'}_{ks} \right) \qquad (k = 1, 2), \quad (8.2\text{-}26)$$

The values obtained from this equation are shown in the first and second columns of the last column of Table 8.2-1. The agreement between corresponding values is excellent.

Actual values of diffusion mass flux j_k, mass transfer coefficient β_{kx}, and Sherwood number Sh_{ks} are defined as

$$j_k = -\rho \sum_{l=1}^{2} D^+_{kl} \left(\frac{\partial W_l}{\partial y} \right)_s = \beta_{kx} (W_{ks} - W_{k\infty})$$

$$(k = 1, 2), \qquad (8.2\text{-}27)$$

$$Sh_{kx} = \frac{\beta_{kx} x}{\mu_V} \qquad (k = 1, 2). \qquad (8.2\text{-}28)$$

From these equations and using Eqs. (7.2-1), (7.2-4), and (8.2-10) we can derive the following equations:

$$Sh_{1x} = -\left\{ \frac{1}{Sc_{11}}\tilde{\Phi}'_{1s} + \frac{(W_{2s} - W_{2\infty})}{Sc_{12}(W_{1s} - W_{1\infty})}\tilde{\Phi}'_{2s} \right\} \frac{Gr_x^{\frac{1}{4}}}{\sqrt{2}}, \qquad (8.2\text{-}29)$$

$$Sh_{2x} = -\left\{ \frac{(W_{1s} - W_{1\infty})}{Sc_{21}(W_{2s} - W_{2\infty})}\tilde{\Phi}'_{1s} + \frac{1}{Sc_{22}}\tilde{\Phi}'_{2s} \right\} \frac{Gr_x^{\frac{1}{4}}}{\sqrt{2}}, \qquad (8.2\text{-}30)$$

where the $\tilde{\Phi}'_{ks}$ are obtained from Eqs. (8.2-26) and (7.3-23).

The local heat transfer coefficient α_{cx} and Nusselt number Nu_{cx} are defined by

$$q_{cx} = \lambda_V \left(\frac{\partial T}{\partial y} \right)_s = \alpha_{cx}(T_{V\infty} - T_s), \qquad (8.2\text{-}31)$$

$$Nu_{cx} = \frac{\alpha_{sx}x}{\lambda_V} = \frac{1}{\sqrt{2}}(-\tilde{\Theta}'_s)Gr_x^{\frac{1}{4}}$$

$$= \frac{1}{\sqrt{2}}(-\Theta'_s)\left(\frac{gx^3}{\nu^2} \right)^{\frac{1}{4}}. \qquad (8.2\text{-}32)$$

Considering the functional form of Eqs. (4.4-5) and (4.4-7), we assume the following correlation equation for Nu_{cx} :

$$Nu_{cx} = C_G(Pr_V)\left\{ \left(\frac{gx_i^+ x^3}{\nu_V^2} \right)Pr_V \right\}^{\frac{1}{4}}, \qquad (8.2\text{-}33)$$

$$\chi_i^+ = \omega_T + \omega_1^* \left(\frac{Pr_V}{Sc_1} \right)^{\frac{1}{2}} + \omega_2^* \left(\frac{Pr_V}{Sc_2} \right)^{\frac{1}{2}} \qquad (8.2\text{-}34\text{a})$$

$$= \frac{1 + N_1 \left(\frac{Pr_V}{Sc_1} \right)^{\frac{1}{2}} + N_2 \left(\frac{Pr_V}{Sc_2} \right)^{\frac{1}{2}}}{1 + N_1 + N_2} \omega_s. \qquad (8.2\text{-}34\text{b})$$

The comparison between Eqs. (8.2-32) and (8.2-33) gives the equation

$$-\Theta'_s = \sqrt{2}C_G(Pr_V)\left(\chi_i^+ Pr_V \right)^{\frac{1}{4}} = \omega_s^{\frac{1}{4}}\left(-\tilde{\Theta}'_s \right). \qquad (8.2\text{-}35)$$

The values obtained from this equation are shown in the last column of Table 8.2-1. The agreement between $\tilde{\Theta}'_s$ and $(\tilde{\Theta}'_s)_{cal}$ is also excellent.

From Eqs. (8.2-25a) and (8.2-34a) we can infer Eqs. (8.1-13) and (8.1-14), respectively.

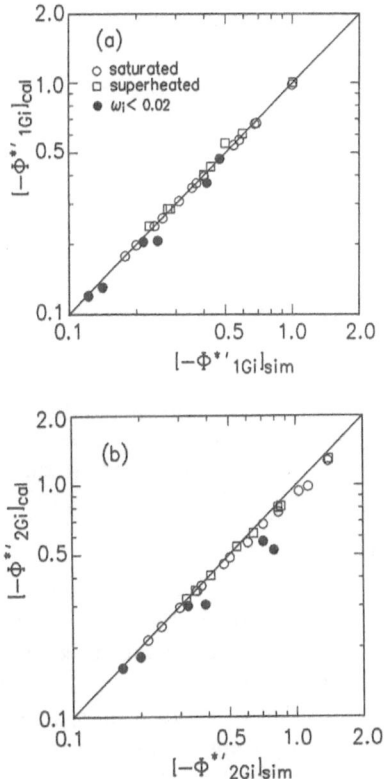

FIGURE 8.3-1.   Comparison of the boundary values of the similarity solution $\Phi^{*'}_{kGi}$ with the values of Eq. (8.1-11) in free-convection condensation of a methanol-ethanol-water mixture. (a) $\Phi^{*'}_{1Gi}$, (b) $\Phi^{*'}_{2Gi}$.

8.3 Free-Convection Condensation of Ternary Vapors [3]

In this case the system of the basic differential equations is given by taking $n = 3$ in the equations in Section 8.1, and the numerical solution can be obtained by the same way as in Section 4.2. By using an example of the results for a methanol-ethanol-water mixture, we now discuss the validity of the algebraic method inferred in Section 8.1.

Figures 8.3-1(a) and (b) show the comparison of the boundary values of the similarity solution $(-\Phi^{*'}_{1Gi})_{sim}$ and $(-\Phi^{*'}_{2Gi})_{sim}$ with the values of Eq. (8.1-11) for $k = 3$, respectively. The agreement between them is within an error of about 5 percent, except the case of $\omega_i < 0.02$ which is denoted by the symbol ●. Figure 8.3-2 shows the comparison of the boundary values of the similarity solution $(-\Theta'_{GVi})_{sim}$ with the values of Eq. (8.1-12) for

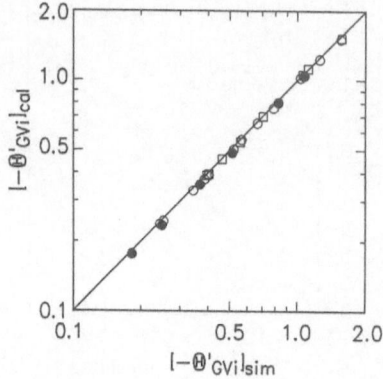

FIGURE 8.3-2. Comparison of the boundary values of the similarity solution Θ'_{GVi} with the values of Eq. (8.1-12) in free-convection condensation of a methanol-ethanol-water mixture. Symbols used are the same as in Fig. 8.3-1.

$k = 3$. The agreement is also good. The boundary values of the similarity solution Θ'_{GLw} and Θ'_{GLi} are correlated well by Eq. (4.4-4).

The algebraic equations for a ternary vapor mixture can be reduced from Eqs. (8.1-16) and (8.1-17). Figures 8.3-3(a) and (b) show the ratio of the algebraic solution to the similarity solution in q_{wx} and $\dot{m}_x$ for a methanol-ethanol-water mixture. The agreement between these solutions is good.

Taitel and Tamir[4] numerically solved the two-phase boundary layer equations for free-convection condensation of air-methanol-water and aceton-methanol-water mixtures, giving constant values of D_{kl}^+ and neglecting

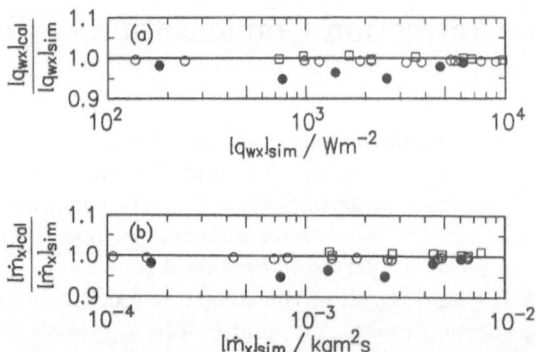

FIGURE 8.3-3. Comparison in q_{wx} and $\dot{m}_x$ between the algebraic and similarity solution in free-convection condensation of a methanol-ethanol-water mixture. Symbols used are the same as in Fig. 8.3-1. (a) $(q_{wx})_{cal}/(q_{wx})_{sim}$, (b) $(\dot{m}_x)_{cal}/(\dot{m}_x)_{sim}$.

buoyancy force based on temperature. They graphically showed the result on heat flux in the coordinate of q_{wx}/q_{wx}^0 versus $T_{V\infty}$ with $(T_{V\infty} - T_w)$ as a parameter, where q_{wx}^0 is the heat flux calculated under the assumption that condensation occurs at bulk conditions. Sage and Estrin[5] obtained the numerical solution for free-convection condensation of nitrogen-neon-water, nitrogen-CFC12-water, and nitrogen-methan-water mixtures, using the method of Toor[6] and Stewart and Prober[7]. They showed the heat transfer characteristics in comparison with those of pure water vapor. They also discussed the appearance of super-saturation in the vapor boundary layer, where the diffusion term in the energy equation which has an important role as shown in Section 4.3.3 is neglected. Fujii et al.[8] confirmed the validity and effectiveness of the above-mentioned present method in the experiment for free-convection condensation of an air-CFC12-CFC114 mixture.

REFERENCES

1. Koyama, Sh., M. Goto, M. Watabe, and T. Fujii, Laminar Film Condensation of Multicomponent Mixtures on a Flat Plate (Second Report, Gravity Controlled Condensation) (in Japanese), *Reports of Institute of Advanced Material Study, Kyushu University*, 1, 1, 85–89 (1987).

2. Watabe, M., T. Fujii, and Sh. Koyama, Simultaneous Heat and Mass Transfer in Laminar Free Convection of Ternary Vapor Mixtures on a Vertical Flat Plate (in Japanese), *Trans. Jpn. Soc. Mech. Eng.*, 54, 499, B, 647–645 (1988).

3. Fujii, T., Sh. Koyama, and M. Watabe, Laminar Film Condensation of Gravity-Controlled Convection for Ternary Vapor Mixtures on a Flat Plate (in Japanese), *Trans. Jpn. Soc. Mech. Eng.*, 55, 510, B, 434–441 (1989).

4. Taitel, Y. and A. Tamir, Film Condensation of Multicomponent Mixtures, *Int. J. Multiphase Flow*, 1, 697–714 (1974).

5. Sage, E. F. and J. Estrin, Film Condensation from a Ternary Mixture of Vapors upon a Vertical Surface, *Int. J. Heat and Mass Transfer*, 19, 323–333 (1976).

6. Toor, H. L., Solution of the Linearized Equations of Multicomponent Mass Transfer: II. Matrix Methods, *AIChE J.*, 10-4, 460–465 (1964).

7. Stewart, W. E. and R. Prober, Matrix Calculation of Multicomponent Mass Transfer in Isothermal Systems, *I & EC Fundament.*, 3-3, 224–235 (1964).

8. Fujii, T., Sh. Koyama, and M. Goto, Effects of Air upon Gravity Controlled Condensation of Nonazeotropic Binary Refrigerant Vapor on a Horizontal Tube (in Japanese), *Trans. Jpn. Soc. Mech. Eng.*, 51, 46, B, 2442–2450 (1985).

9

Representative Physical Properties for the Condensate Film and the Vapor Boundary Layer

Physical properties vary with temperature and concentration. In the previous theoretical analyses, constant property values have been given under the assumption that representative values will be adequate. These will be determined in this chapter. The variation of physical properties in the condensate film is smaller than that in the vapor boundary layer. Therefore, representative values for the condensate film are first determined by calculation for saturated pure vapors. Those for the vapor boundary layer of binary vapor mixtures are then determined by calculation, considering the distribution of both temperature and concentration, while using the obtained representative values for the condensate film.

9.1 Representative Physical Properties for the Condensate Film

As discussed in Chapters 3 to 5, the temperature distribution in the condensate film is nearly linear in both forced- and free-convection condensation. Among the relevant physical properties, the variation of viscosity is largest and the variations of the other physical properties are smaller in the order of thermal conductivity or specific heat and density, except special cases as shown in Chapter 10. When thermal conductivity varies linearly with temperature in the case of one-dimensional conductive heat transfers in a solid, it is easily shown that the representative thermal conductivity for calculating the heat flux is that at average temperature. In the present problem the variation of viscosity affects the film thickness, and accordingly the condensation heat transfer coefficient. This characteristic is first considered by an approximate analysis and then by a numerical calculation.

9.1.1   APPROXIMATE ANALYSIS FOR FORCED-CONVECTION CONDENSATION OF A SATURATED PURE VAPOR[1]

We assume that the temperature distribution in the condensate film is linear and only the viscosity varies with temperature. Accordingly, the heat flux q_{wx} and the dynamic viscosity $\mu_L(T^*)$ are expressed as

$$q_{wx} = \lambda_L \frac{T_s^*}{\delta} = \lambda_L \frac{T^*}{y} , \qquad (9.1\text{-}1a,b)$$

$$\frac{1}{\mu_L(T^*)} = \frac{1}{\mu_{Lw}} (1 + AT^*) , \qquad (9.1\text{-}2)$$

where λ_L is the thermal conductivity which is evaluated at $(T_w + T_s)/2$, δ is the film thickness, $T^* = T_L - T_w$ is the temperature relative to T_w, μ_{Lw} is the dynamic viscosity at the cooling surface, and A is a constant peculiar to relevant substance. From Eqs. (9.1-1) and (9.1-2) the following equation is derived :

$$\frac{1}{\mu_L(T^*)} = \frac{1}{\mu_{Lw}} \left(1 + \frac{Aq_{wx}}{\lambda_L} y \right) = \frac{1}{\mu_{Lw}} (1 + B_x y) , \qquad (9.1\text{-}3a,b)$$

where

$$B_x = \frac{Aq_{wx}}{\lambda_L} = \frac{AT_s^*}{\delta} . \qquad (9.1\text{-}4a,b)$$

We use the approximate method as explained in Section 5.6, i.e., we express the shear stress at the vapor-liquid interface as

$$\tau_w = \tau_i = \mu_L(T^*) \left(\frac{\partial U_L}{\partial y} \right)_i = 4\chi^2 \dot{m}_x U_{V\infty} . \qquad (9.1\text{-}5a,b)$$

For large values of RH/Pr_L, i.e., for thick condensate films, which correspond to $\dot{M}_{FV} \to \infty$, χ of Eq. (5.1-9) asymptotically approaches a constant value 0.478. Substituting this value and $\mu_L(T^*)$ of Eq. (9.1-3) into Eq. (9.1-5), we obtain

$$\frac{\partial U_L}{\partial y} = \frac{4 \times 0.229}{\mu_{Lw}} (1 + B_x y) \dot{m}_x U_{V\infty} = K\dot{m}_x (1 + B_x y) , \qquad (9.1\text{-}6)$$

where

$$K = \frac{4 \times 0.229 U_{V\infty}}{\mu_{Lw}} . \qquad (9.1\text{-}7)$$

Equation (9.1-6) is solved under the condition of $U_L = 0$ at $y = 0$ to give

$$U_L = K\dot{m}_x \left(y + \frac{B_x}{2} y^2 \right) . \qquad (9.1\text{-}8)$$

Substituting Eq. (9.1-8) into Eq. (2-35), and eliminating $\dot{m}_x$ by using the equation

$$\dot{m}_x = \frac{q_{wx}}{\Delta h_V} = \frac{\lambda_L T_s^{\,*}}{\Delta h_V \delta} \tag{9.1-9}$$

and using B_x of Eq. (9.1-4b), we obtain

$$\frac{d\delta^2}{dx} = \frac{4}{K\rho_L}\left(1 + \frac{AT_s^{\,*}}{3}\right)^{-1}. \tag{9.1-10}$$

Equation (9.1-10) is solved under the condition of $\delta=0$ at $x=0$ to give

$$\delta^2 = \frac{4}{K\rho_L}\left(1 + \frac{AT_s^{\,*}}{3}\right)^{-1} x. \tag{9.1-11}$$

The following equation is consequently obtained:

$$Nu_{Lwx} = \frac{x}{\delta} = 0.479\left(1 + \frac{AT_s^{\,*}}{3}\right)^{\frac{1}{2}}\left(\frac{\rho_L U_{V\infty} x}{\mu_{Lw}}\right)^{\frac{1}{2}}. \tag{9.1-12}$$

We denote the film thickness δ_r and the dynamic viscosity μ_{Lr} for the case of $A = 0$, i.e., constant μ_L. When q_{wx} for variable viscosity is equal to that for constant viscosity for a given value of T_s^*, δ must be equal to δ_r. Then, Eq. (9.1-12) becomes

$$(Nu_{Lwx})_r = \frac{x}{\delta_r} = 0.479\left(\frac{\rho_L U_{V\infty} x}{\mu_{Lr}}\right)^{\frac{1}{2}}. \tag{9.1-13}$$

From the relation $\delta = \delta_r$ or $Nu_{Lwx} = (Nu_{Lwx})_r$, the following equation is derived:

$$\frac{\mu_{Lw}}{\mu_{Lr}} = \left(1 + \frac{AT_s^{\,*}}{3}\right). \tag{9.1-14}$$

Furthermore, the following equation is derived from Eq. (9.1-2) by denoting T_r^* as the temperature corresponding to μ_{Lr}:

$$T_r^* = \frac{1}{3}T_s^{\,*} \tag{9.1-15}$$

or

$$r_T = \frac{T_{Lr} - T_w}{T_s - T_w} = \frac{1}{3}, \tag{9.1-16}$$

where r_T denotes the dimensionless temperature for representative physical properties.

For small values of RH/Pr_L, i.e., for thin condensate films, which correspond to $\dot{M}_{FV} \to 0$, Eq. (5.1-9) can be approximated as

$$\chi \approx 0.45 \left(\frac{Pr_{Lw}}{R_w H}\right)^{\frac{1}{3}} \left(\frac{\mu_L}{\mu_{Lw}}\right)^{\frac{1}{6}}. \tag{9.1-17}$$

As in the case of large values of RH/Pr_L mentioned above, we can derive the following equations:

$$Nu_{Lwx} = 0.45 \left(\frac{Pr_{Lw}}{R_w H}\right)^{\frac{1}{3}}$$

$$\times \left\{1 + \frac{2}{9}AT_s{}^* - \frac{1}{54}(AT_s{}^*)^2 + \cdots\right\}^{\frac{1}{2}} \left(\frac{\rho_L U_{V\infty} x}{\mu_{Lw}}\right)^{\frac{1}{2}}, \tag{9.1-18}$$

$$\frac{\mu_{Lw}}{\mu_{Lr}} \approx \left\{1 + \frac{2}{9}AT_s{}^* - \frac{1}{54}(AT_s{}^*)^2 + \cdots\right\}^{\frac{3}{2}}, \tag{9.1-19}$$

$$r_T = \frac{T_{Lr} - T_w}{T_s - T_w} = \frac{1}{3}\left(1 - \frac{\mu_{Lw} - \mu_{Ls}}{36\mu_{Ls}}\right). \tag{9.1-20}$$

Since the value of $(\mu_{Lw} - \mu_{Ls})/\mu_{Ls}$ in Eq. (9.1-20) is estimated to be of the order of one at most, the effect of the second term in Eq. (9.1-20) is small. Therefore, the dimensionless representative temperature for forced-convection condensation can be regarded as $r_T = 1/3$. Note that this value has been obtained based on Eq. (5.1-8).

9.1.2   NUMERICAL ANALYSIS FOR FORCED-CONVECTION CONDENSATION OF SATURATED PURE VAPORS[2]

For constant physical properties except dynamic viscosity, the basic equations are written as

$$\left(\frac{\mu_L}{\mu_{Ls}}F''_{FL}\right)' + \frac{1}{2}F_{FL}F''_{FL} = 0, \tag{9.1-21}$$

$$\Theta''_{FL} + \frac{Pr_{Ls}}{2}F_{FL}\Theta'_{FL} = 0, \tag{9.1-22}$$

$$F'''_{FV} + \frac{1}{2}F_{FV}F''_{FV} = 0, \tag{9.1-23}$$

together with the boundary conditions

at $\eta_{FL} = 0$: $F_{FLw} = 0$, $F'_{FLw} = 0$, $\Theta_{FLw} = 1$; \hfill (9.1-24)

as $\eta_{FV} \to \infty$: $F'_{FV\infty} = 1$; \hfill (9.1-25)

and with the compatibility conditions at the vapor-liquid interface:

at $\eta_{FL} = \eta_{FLi}$ or $\eta_{FV} = 0$:

$$F_{FVi} = \left(\frac{\rho_L \mu_{Ls}}{\rho_{Vs}\mu_{Vs}}\right)^{\frac{1}{2}} F_{FLi} , \qquad (9.1\text{-}26)$$

$$F'_{FVi} = F'_{FLi} , \qquad (9.1\text{-}27)$$

$$F''_{FVi} = \left(\frac{\rho_L \mu_{Ls}}{\rho_{Vs}\mu_{Vs}}\right)^{\frac{1}{2}} F''_{FLi} , \qquad (9.1\text{-}28)$$

$$\Theta_{FLi} = 0 , \qquad (9.1\text{-}29)$$

$$-\Theta'_{FVi} = \left(\frac{\mu_{Ls}}{\lambda_L}\right)\frac{\Delta h_V \dot{M}_{FLs}}{(T_s - T_w)} , \qquad (9.1\text{-}30)$$

where

$$\dot{M}_{FLs} = \frac{\dot{m}_x x}{\mu_{Ls}}\left(\frac{U_{V\infty}x}{\nu_{Ls}}\right)^{-\frac{1}{2}} . \qquad (9.1\text{-}31)$$

Equations (9.1-21) to (9.1-30) can be solved numerically for a given fluid by using the relation between μ_L and T_L, giving the values of λ_L, c_{pL}, and ρ_L at $(T_w + T_s)/2$ and the values of μ_V, ρ_V, and Δh_V at T_s. Then the heat transfer coefficient α_{wx} and the condensation mass flux $\dot{m}_x$ are obtained by

$$\alpha_{wx}\left(\frac{x}{U_{V\infty}}\right)^{\frac{1}{2}} = \lambda_L\left(\frac{\rho_L}{\mu_{Ls}}\right)^{\frac{1}{2}}(-\Theta'_{FLw}) , \qquad (9.1\text{-}32)$$

$$\dot{m}_x\left(\frac{x}{U_{V\infty}}\right)^{\frac{1}{2}} = (\mu_{Ls}\rho_L)^{\frac{1}{2}}\frac{F_{FLi}}{2} . \qquad (9.1\text{-}33)$$

On the other hand, for the constant value of μ_{Le} at T_e, the heat transfer coefficient $(\alpha_{wx})_e$, which is obtained from Eq. (5.1-4) or (5.1-8), and the condensation mass flux $(\dot{m}_x)_e$, which is obtained from the solution of Eq. (5.1-1), are expressed as

$$(\alpha_{wx})_e\left(\frac{x}{U_{V\infty}}\right)^{\frac{1}{2}} = 0.433\left\{1.367 - \frac{0.432}{\sqrt{(2\dot{M}_{FV})_e}} + \frac{1}{(2\dot{M}_{FV})_e}\right\}^{\frac{1}{2}}\frac{\lambda_L\sqrt{\rho_L}}{\sqrt{\mu_{Le}}} \tag{9.1-34a}$$

$$= 0.45\left\{1.20 + \frac{Pr_{Le}}{(R)_e H}\right\}^{\frac{1}{3}}\frac{\lambda_L\sqrt{\rho_L}}{\sqrt{\mu_{Le}}} , \qquad (9.1\text{-}34\text{b})$$

$$(\dot{m}_x)_e\left(\frac{x}{U_{V\infty}}\right)^{\frac{1}{2}} = (\rho_V\mu_V)^{\frac{1}{2}}\left(\dot{M}_{FV}\right)_e , \qquad (9.1\text{-}35)$$

respectively.

Figures 9.1-1 and 9.1-2 show the result for saturated steam of $T_s = 30°C$

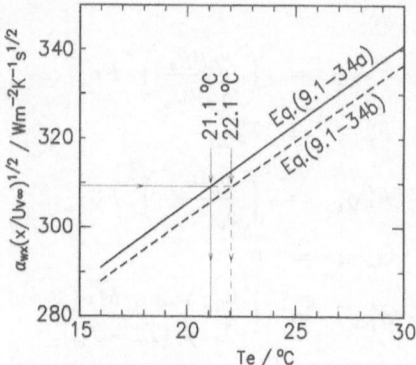

FIGURE 9.1-1.   Comparison of the value of $\alpha_{wx}(x/U_{V\infty})^{1/2}$ from Eq. (9.1-32)(numerical solution) with those of $(\alpha_{wx})_e(x/U_{V\infty})^{1/2}$ from Eqs. (9.1-34a) and (9.1-34b)(algebraic solution), in which the relevant physical properties are evaluated at T_e, in the case of forced-convection condensation of saturated steam at $T_s = 30°C$ and $T_w = 15.94°C$.

and $T_w = 15.94°C$ as an example. The physical properties used are shown in the Appendix. The solid and dashed lines in Fig. 9.1-1 represent the values of $(\alpha_{wx})_e(x/U_{V\infty})^{1/2}$ which are obtained by substituting the physical properties at T_e into Eqs. (9.1-34a) and (9.1-34b), respectively. Since the value of $\alpha_{wx}(x/U_{V\infty})^{1/2}$ in Eq. (9.1-32) for the numerical results is 309.3 $Wm^{-2}K^{-1}s^{1/2}$, we can obtain the representative temperatures $T_{Lr} = 21.1°C$ for Eq. (9.1-34a) and $T_{Lr} = 21.1°C$ for Eq. (9.1-34b) or $r_T = 0.37$ and 0.44, respectively. The difference between the numerical

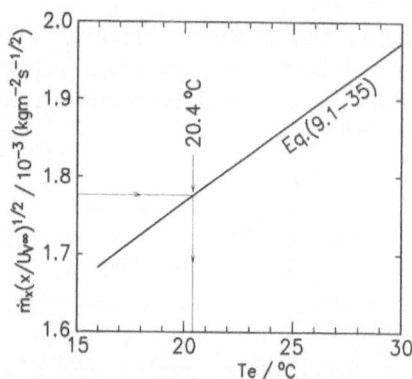

FIGURE 9.1-2. Comparison of the value of $\dot{m}_x(x/U_{V\infty})^{1/2}$ from Eq. (9.1-33) (numerical solution) with those of $(\dot{m}_x)_e(x/U_{V\infty})^{1/2}$ from Eqs. (9.1-35) (algebraic solution), in which the relevant physical properties are evaluated at T_e, in the case of forced-convection condensation of saturated steam at $T_s = 30°C$ and $T_w = 15.94°C$.

results α_{wx} and the value $(\alpha_{wx})_e$ at $T_e = T_w$ and that between α_{wx} and $(\alpha_{wx})_e$ at $T_e = T_s$ are about -8 percent and $+13$ percent, respectively, while the difference between α_{wx} and $(\alpha_{wx})_e$ at T_e which corresponds to $r_T = 1/3$ is only about -2 percent. The difference between the values of T_{Lr} and r_T based on Eq. (9.1-34a) and those based on Eq. (9.1-34b) is caused by the difference in accuracy between these equations.

Figure 9.1-2 shows the comparison between Eq. (9.1-35) and (9.1-33) obtained as above. Corresponding to the numerical result $\dot{m}_x(x/U_{V\infty})^{1/2} = 1.776 \times 10^{-3}$ kg m^{-2}s$^{-1/2}$ in Eq. (9.1-33), $T_{Lr} = 20.4°\text{C}$ and $r_T = 0.32$ are obtained. These values are almost the same as those for the approximate analytical solution.

Table 9.1-1 shows a few examples of the numerical results of α_{wx} and $\dot{m}_x$ and the values of r_T, $(\alpha_{wx})_e$ and $(\dot{m}_x)_e$ which are evaluated as mentioned above for the saturated vapors of water, ethanol, ethylene glycol, and propane. In the calculation the physical properties are referred from the equations and references in the Appendix. Although the value of r_T is somewhat irregular, $r_T = 1/3$ which is derived by the approximate analysis in Section 9.1.1 is practically recommendable for each substance. The same conclusion was also derived from the numerical calculation by Fujii et al.,[2] in which the numerical solutions taken into account the variation of all relevant physical properties are compared with those for constant physical properties.

As already mentioned, the value of r_T is somewhat affected by the correlation equation for Nu_{Lwx} and $\dot{M}_{FV}$ as well as by the values of RH/Pr_L. Sparrow et al.[3] deemed it sufficient to use $r_T = 1/3$ for steam by referring to the results for the case of free-convection condensation. Denny and Mills[4] made a numerical analysis of the combined forced- and free-convection condensation as a variable property problem, and obtained the following results: $r_T = 0.07, 0.12, 0.15, 0.25, 0.29, 0.29, 0.32, 0.33, 0.61,$ and 1.00 for carbon tetrachloride, ethanol, n-propyl alcohol, n-butyl alcohol, t-butyl alcohol, ethylene glycol, glyserol, water, ammonia, and propane, respectively. Such variety of the r_T value is inappropriate from the viewpoint of the physical meaning of the representative value. This is considered to be due to the fact that their numerical results are incorrect and the equation of the Nusselt number used for the reference is inadequate, because they used the equation of Shekriladze and Gomelauri[5] for the shear stress at the vapor-liquid interface (see Section 5.6).

9.1.3 APPROXIMATE AND NUMERICAL ANALYSES FOR FREE-CONVECTION CONDENSATION OF SATURATED PURE VAPORS

Under the same assumptions as the case of forced-convection condensation, and neglecting the shear stress at the vapor-liquid interface and the inertia

TABLE 9.1-1. Examples of the numerical results of α_{wx}, $\dot{m}_x$, and r_T, where only the variation of viscosity is taken into account. The values of $(\alpha_{wx})_e$ and $(\dot{m}_x)_e$ for T_e corresponding to $r_T=0$, 1/3, and 1 are also presented for reference.

			1. water*	2. water*	3. ethanol	4. ethylene-glycol*	5. propane
T_s/°C			30	100	40	60	30
T_w/°C			15.94	71.97	19.06	46.61	14.13
$\alpha_{wx}(x/U_{V\infty})^{1/2}$/Wm^{-2}K^{-1}s$^{1/2}$			309.3	568.3	76.5	53.3	125.4
r_T		Eq. (9.1-34a)	0.37	0.31	0.34	0.37	0.21
		Eq. (9.1-34b)	0.44	0.36	0.35	0.43	0.32
$(\alpha_{wx})_e(x/U_{V\infty})^{1/2}$ Wm^{-2}K^{-1}s$^{1/2}$	T_e corr. to r_T=1/3	Eq. (9.1-34a)	308	570	76.4	52.9	126.9
		Eq. (9.1-34b)	304	565	76.3	52.3	125.5
	$T_e = T_w$ ($r_T = 0$)	Eq. (9.1-34a)	290	537	71.5	49.3	123.0
		Eq. (9.1-34b)	288	534	71.9	48.9	122.3
	$T_e = T_s$ ($r_T = 1$)	Eq. (9.1-34a)	342	634	86.7	60.3	135.3
		Eq. (9.1-34b)	337	626	85.5	59.3	132.2
$\dot{m}_x(x/U_{V\infty})^{1/2}$/kgm^{-2}s$^{-1/2}$			1.7762×10^{-3}	6.9461×10^{-3}	1.7431×10^{-3}	6.6808×10^{-4}	5.8000×10^{-3}
r_T			0.32	0.24	0.36	0.32	0.22
$(\dot{m}_x)_e(x/U_{V\infty})^{1/2}$ kgm^{-2}s$^{-1/2}$	T_e corr. to $r_T = 1/3$		1.7800×10^{-3}	7.0558×10^{-3}	1.7359×10^{-3}	6.7011×10^{-4}	5.8521×10^{-3}
	$T_e = T_w$ ($r_T = 0$)		1.6826×10^{-3}	6.6569×10^{-3}	1.6356×10^{-3}	6.2577×10^{-4}	5.7048×10^{-3}
	$T_e = T_s$ ($r_T = 1$)		1.9736×10^{-3}	7.8204×10^{-3}	1.9472×10^{-3}	7.6166×10^{-4}	6.1650×10^{-3}

* In the case of solving $\dot{M}_{FV}$ from Eq. (5.1-1), the correction due to convection term in Eq. (5.1-1) is not necessary.

term in the condensate film, the following basic equation is derived:

$$\frac{\partial}{\partial y}\left\{\mu_L(T^*)\frac{\partial U_L}{\partial y}\right\} = -\rho_L g \tag{9.1-36}$$

and by using the same procedure as the forced-convection condensation, the following results for Nu_{Lwx}, μ_{Lr}, and r_T can be obtained:

$$Nu_{Lwx} = \frac{1}{\sqrt{2}}\left(1+\frac{AT_s^{\,*}}{4}\right)^{\frac{1}{4}}\left(\frac{Ga_{wx}Pr_{Lw}}{H}\right)^{\frac{1}{4}}, \tag{9.1-37}$$

$$\frac{\mu_{Lw}}{\mu_{Lr}} = 1+\frac{AT_s^{\,*}}{4}, \tag{9.1-38}$$

$$r_T = \frac{T_{Lr}-T_w}{T_s-T_w} = \frac{1}{4}. \tag{9.1-39}$$

Drew, as described by McAdams,[6] found the same result as Eq. (9.1-39). Minkowycz and Sparrow[7] derived r_T=0.31 from the numerical results, which were obtained under the same assumptions as above for steam of $T_s = 22$–$100°C$ and for $T_s-T_w = 1$–$25°C$. Poots and Miles[8] solved the same equations as those in Chapter 10 for steam of $T_s = 100°C$ and wall temperature of 0–$90°C$, compared their results with those of Voskresenskiy,[9] Labuntsov,[10] and Minkowycz and Sparrow, and proposed the following equation as the representative temperature for Nusselt's equation (5.3-2):

$$r_T = \frac{T_{Lr}-T_w}{T_s-T_w} = 0.2474 + 0.1580\left(\frac{T_s-T_w}{T_s}\right) - 0.0769\left(\frac{T_s-T_w}{T_s}\right)^2, \tag{9.1-40}$$

where T denotes Celsius temperature. The value of r_T given by this equation is nearly equal to $1/4$ when $(T_s - T_w)$ becomes zero, and does not increase so much with $(T_s - T_w)$. The difference between the wall heat flux corresponding to $r_T = 0.31$ and that corresponding to $r_T = 1/4$ is only one percent.

9.2 Representative Physical Properties for the Vapor Boundary Layer

The basic equations in the vapor boundary layer are written as

$$\frac{\partial(\rho_V U_V)}{\partial x} + \frac{\partial(\rho_V V_V)}{\partial y} = 0, \tag{9.2-1}$$

$$\rho_V \left(U_V \frac{\partial U_V}{\partial x} + V_V \frac{\partial U_V}{\partial y} \right) = \frac{\partial}{\partial y}\left(\mu_V \frac{\partial U_V}{\partial y} \right) + g(\rho_V - \rho_{V\infty}), \tag{9.2-2}$$

$$\rho_V c_{pV} \left(U_V \frac{\partial T_V}{\partial x} + V_V \frac{\partial T_V}{\partial y} \right) = \frac{\partial}{\partial y}\left(\lambda_V \frac{\partial T_V}{\partial y} \right), \tag{9.2-3}$$

$$\rho_V \left(U_V \frac{\partial W_{1V}}{\partial x} + V_V \frac{\partial W_{1V}}{\partial y} \right) = \frac{\partial}{\partial y}\left(\rho_V D \frac{\partial W_{1V}}{\partial y} \right). \tag{9.2-4}$$

The diffusion term in Eq. (9.2-3) has been neglected because its effect on the heat flux at the cooling surface and condensation mass flux is negligibly small. Also, the second term on the right-hand side of Eq. (9.2-2) is ignored for forced-convection condensation.

Equations (2-1) to (2-3) are used for the basic equations of the condensate film, and the relevant boundary and compatibility conditions are given by Eqs. (2-14) to (2-19) and Eqs. (2-20) to (2-27), respectively.

9.2.1 FORCED-CONVECTION CONDENSATION OF BINARY VAPORS[11]

The stream function and the variables of the similarity transformation are defined as

$$U_V = \frac{\rho_{V\infty}}{\rho_V} \frac{\partial \Psi_V}{\partial y}, \tag{9.2-5a}$$

$$V_V = -\frac{\rho_{V\infty}}{\rho_V} \frac{\partial \Psi_V}{\partial x}, \tag{9.2-5b}$$

$$\eta_{FV} = \left(\frac{U_{V\infty}}{\nu_{V\infty} x} \right)^{\frac{1}{2}} \int_{\delta}^{y} \frac{\rho_V}{\rho_{V\infty}} dy, \tag{9.2-6}$$

$$F_{FV} = \frac{\Psi_V}{(\nu_{V\infty} U_{V\infty} x)^{\frac{1}{2}}}. \tag{9.2-7}$$

Θ_{FV} and Φ_{FV} are defined by Eqs. (3.1-8) and (3.1-9), respectively.

The transformation of the basic equations together with the boundary and compatibility conditions then yields

$$\left(\frac{\rho_V \mu_V}{\rho_{V\infty} \mu_{V\infty}} F''_{FV} \right)' + \frac{1}{2} F_{FV} F''_{FV} = 0, \tag{9.2-8}$$

$$\left(\frac{\rho_V \lambda_V}{\rho_{V\infty} \lambda_{V\infty}} \Theta'_{FV}\right)' + \frac{1}{2} \frac{c_{pV} \mu_{V\infty}}{\lambda_{V\infty}} F_{FV} \Theta'_{FV} = 0, \qquad (9.2\text{-}9)$$

$$\left(\frac{\rho_V^2 D}{\rho_{V\infty}^2 D_\infty} \Phi'_V\right)' + \frac{1}{2} Sc_\infty F_{FV} \Phi'_{FV} = 0, \qquad (9.2\text{-}10)$$

$$F''_{FL} + \frac{1}{2} F_{FL} F''_{FL} = 0, \qquad (9.2\text{-}11)$$

$$\Theta''_{FL} + \frac{1}{2} Pr_L F_{FL} \Theta'_{FL} = 0; \qquad (9.2\text{-}12)$$

at $\eta_{FL} = 0$:

$$F_{FLw} = 0, \qquad (9.2\text{-}13)$$

$$F'_{FLw} = 0, \qquad (9.2\text{-}14)$$

$$\Theta_{FLw} = 1; \qquad (9.2\text{-}15)$$

as $\eta_{FV} \to \infty$:

$$F'_{FL\infty} = 1, \qquad (9.2\text{-}16)$$

$$\Theta_{FV\infty} = 0, \qquad (9.2\text{-}17)$$

$$\Phi_{FV\infty} = 0; \qquad (9.2\text{-}18)$$

at $\eta_{FL} = \eta_{FLi}$ or $\eta_{FV} = 0$:

$$F_{FVi} = R_\infty F_{FLi} = 2R_\infty \dot{M}_{FL}, \qquad (9.2\text{-}19)$$

$$F'_{FVi} = F'_{FLi}, \qquad (9.2\text{-}20)$$

$$F''_{FVi}, = \left(\frac{\rho_{V\infty} \mu_{V\infty}}{\rho_{Vi} \mu_{Vi}}\right) R_i F''_{FLi}, \qquad (9.2\text{-}21)$$

$$\Theta_{FLi} = 0, \qquad (9.2\text{-}22)$$

$$\Theta_{FVi} = 1, \qquad (9.2\text{-}23)$$

$$-\Theta'_{FLi} = \frac{Pr_L \dot{M}_{FL}}{H_i} + \frac{\rho_{Vi} \lambda_{Vi}}{\rho_{V\infty} \lambda_L} \left(\frac{\nu_L}{\nu_{V\infty}}\right)^{\frac{1}{2}} \frac{(T_{V\infty} - T_i)}{(T_i - T_w)} (-\Theta'_{FVi}), \qquad (9.2\text{-}24)$$

$$\Phi_{FVi} = 1, \qquad (9.2\text{-}25)$$

$$1 - \left(\frac{\rho_{V\infty}^2 D_\infty}{\rho_{Vi}^2 D_i}\right) \frac{R_\infty Sc_\infty \dot{M}_{FL}}{(-\Phi'_{FVi})} = \frac{W_{1V\infty} - W_{1L}}{W_{1Vi} - W_{1L}}. \qquad (9.2\text{-}26)$$

The heat flux q_{wx} at the cooling surface and the condensation mass flux $\dot{m}_x$ are obtained by

$$q_{wx} \left(\frac{x}{U_{V\infty}}\right)^{\frac{1}{2}} = \frac{\lambda_L}{\sqrt{\nu_L}} (-\Theta'_{FLw})(T_i - T_w), \qquad (9.2\text{-}27)$$

$$\dot{m}_x \left(\frac{x}{U_{V\infty}}\right)^{\frac{1}{2}} = (\rho_L \mu_L)^{\frac{1}{2}} \frac{F_{FLi}}{2}, \qquad (9.2\text{-}28)$$

where Θ'_{FLw} and F_{FLi} are the boundary values of the numerical solution of Eqs. (9.2-8) to (9.2-26).

The heat flux $(q_{wx})_e$ and the condensation mass flux $(\dot{m}_x)_e$, which are obtained from the simultaneous algebraic equations in Chapter 3, are expressed as

$$(q_{wx})_e \left(\frac{x}{U_{V\infty}} \right)^{\frac{1}{2}} = \frac{\lambda_L}{\sqrt{\nu_L}}$$

$$\times \left\{ 0.433 \left(1.367 - \frac{0.432}{\sqrt{2(\dot{M}_{FV})_e}} + \frac{1}{2(\dot{M}_{FV})_e} \right)^{\frac{1}{2}} \right\} (T_i - T_w) , \quad (9.2\text{-}29)$$

$$(\dot{m}_x)_e \left(\frac{x}{U_{V\infty}} \right)^{\frac{1}{2}} = (\rho_L \mu_L)^{\frac{1}{2}} \left(\frac{\dot{M}_{FV}}{R} \right)_e = (\rho_V \mu_V)^{\frac{1}{2}} (\dot{M}_{FV})_e. \quad (9.2\text{-}30)$$

By comparing Eq. (9.2-27) with Eq. (9.2-29) and/or Eq. (9.2-28) with Eq. (9.2-30), we can obtain the representative temperature T_{Vr} and representative concentration W_{1Vr} for the representative physical properties of the vapor boundary layer. These are used in the algebraic solution to give the closest agreement with the numerical solution.

Table 9.2-1 shows some examples of the boundary values, heat flux at the wall and condensation mass flux of the numerical solution for air-water and ethanol-water mixtures. The comparison in the values of $q_{wx}(x/U_{V\infty})^{1/2}$ and $\dot{m}_x(x/U_{V\infty})^{1/2}$ between the numerical and algebraic solutions are shown in Figs. 9.2-1(a), (b), and (c), which correspond to the cases of AW-1, AW-2, and EW-1 in Table 9.2-1, respectively. The physical properties for the algebraic solution are taken at $W_{1V\infty}$ (bulk vapor mixture), W_{1Vi} (vapor-liquid interface), and $(W_{1V\infty} + W_{1Vi})/2$ (saturated), which are pointed by arrows in the abscissa of each figure. In the figures, the symbols □ and ○ represent the values of $(q_{wx})_e (x/U_{V\infty})^{1/2}$ given by Eq. (9.2-29) and $(\dot{m}_x)_e (x/U_{V\infty})^{1/2}$ given by Eq. (9.2-30), respectively. The symbols ■ and ● represent the results for degree of superheat of 100°C corresponding to the conditions of □ and ○, respectively, and horizontal lines, —— (saturated) and – – – (superheated), represent the numerical results shown in Table 9.2-1.

In the case of saturated vapor, the values of $(q_{wx})_e$ and $(\dot{m}_x)_e$ which are obtained by using the physical properties at arithmetic mean concentration and its saturation temperature are close to the respective numerical results. $(q_{wx})_e$ and $(\dot{m}_x)_e$, which are obtained by using the physical properties at arithmetic mean concentration and arithmetic mean temperature (between the bulk and vapor-liquid interface), are also close to the respective numerical results, although they are not shown in the figures. The arithmetic mean values of $(q_{wx})_e$ and $(\dot{m}_x)_e$ which are evaluated by using physical properties at the bulk and vapor-liquid inteface are also close to the re-

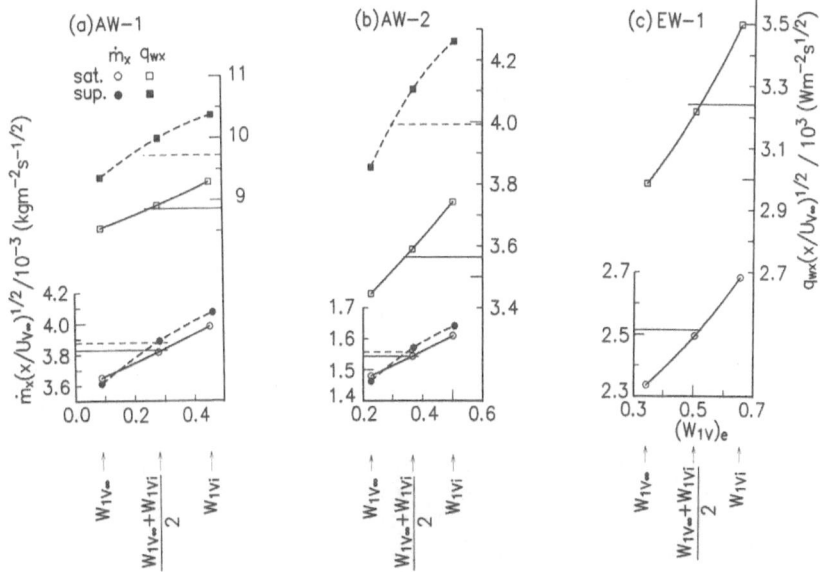

FIGURE 9.2-1.   Comparison of the algebraic solutions $(q_{wx})_e(x/U_{V\infty})^{1/2}$ from Eq.   (9.2-29) ($\square$ —$\square$:   saturated, $\blacksquare$ —$\blacksquare$:   superheated) and $(\dot{m}_x)_e(x/U_{V\infty})^{1/2}$ from Eq. (9.2-30) ($\bigcirc$—$\bigcirc$: saturated, $\bullet$—$\bullet$: superheated) with the corresponding numerical solutions $q_{wx}(x/U_{V\infty})^{1/2}$ from Eq. (9.2-27) and $\dot{m}_x(x/U_{V\infty})^{1/2}$ from Eq. (9.2-28)(——— : saturated, - - - - - : superheated) in the case of forced-convection condensation of binary vapor mixtures. (a) Air-water (AW-1Sa and AW-1Su in Table 9.2-1), (b) air-water (AW-2Sa and AW-2Su in Table 9.2-1), (c) ethanol-water (EW-1Sa in Table 9.2-1).

TABLE 9.2-1. Numerical results for forced-convection condensation of binary vapor mixtures ($p = 0.1$ MPa) in the case where the physical properties in the vapor boundary layer depend on concentration and temperature. AW: air-water; EW: ethanol-water; Sa: saturated; Su: superheated.

No.	$T_{V\infty}$ °C	$W_{1V\infty}$	η_{FLi}	$\dfrac{F''_{FLw}}{10^{-3}}$	F_{FVi}	$\dfrac{F'_{FVi}}{10^{-3}}$	F''_{FVi}	$-\Phi'_{Fi}$	$-\Theta'_{FLw}$	W_{1Vi}	$\dfrac{T_i}{°C}$	$\dfrac{T_w}{°C}$	$q_{wx}(x/U_{V\infty})^{1/2}$ $\mathrm{Wm^{-2}s^{1/2}}$ 10^3	$\dot m_x(x/U_{V\infty})^{1/2}$ $\mathrm{kgm^{-2}s^{-1/2}}$ 10^{-3}
AW-1Sa	98.0	0.0886	1.9	7.1268	2.7711	13.527	1.0667	0.6461	0.5275	0.4713	87.76	72.39	8.859	3.828
AW-2Sa	95.0	0.2275	1.65	3.9448	1.0499	6.507	0.5853	0.3944	0.6065	0.5076	86.36	81.18	3.564	1.5428
EW-1Sa	95.0	0.3428	1.80	4.6175	1.6970	8.307	0.8075	0.4614	0.5568	0.6656	86.20	71.93	5.094	2.516
AW-1Su	198.0	0.0886	1.9	7.1262	2.7487	13.526	1.0812	0.6507	0.5276	0.4905	87.03	69.93	9.724	3.877
AW-2Su	195.0	0.2275	1.65	3.9592	1.0434	6.530	0.5937	0.3949	0.6065	0.5167	85.99	80.16	3.992	1.5565
AW-3Su	198.0	0.0886	1.8908	6.8709	2.5890	12.979	1.0620	0.6398	0.5300	0.4595	88.19	72.39	9.161	3.652
AW-4Su	195.0	0.2275	1.6333	3.8625	0.9916	6.306	0.5858	0.3910	0.6127	0.5004	86.65	81.18	3.802	1.4792

spective numerical results. Consequently, to obtain the algebraic solution we may choose either method as above. Incidentally, the heat transfer coefficient $(\alpha_{wx})_e$ is almost the same for every calculation using different values of $(W_{1V})_e$. However, the values of $(T_i - T_w)$ are different and accordingly, the $(q_{wx})_e$ values also differ.

In the case of superheated vapors, the relation between the numerical and algebraic solutions for $\dot{m}_x$ is almost the same as in the case of saturated vapors. However, q_{wx} of the algebraic solution, which is obtained by using the physical properties at $(W_{1V\infty} + W_{1Vi})/2$ and the corresponding saturation temperature, is higher than that of the corresponding numerical solution by about three percent. It seems to be caused by the accuracy of algebraic solution (see Section 3.6). Therefore, we can conclude that the same method for choosing representative values is valid for saturated and superheated vapors.

9.2.2 FREE-CONVECTION CONDENSATION OF BINARY VAPORS[12]

The stream function is the same as Eq. (9.2-5) and the similarity variables are taken as

$$\eta_{GV} = \left(\frac{g}{4\nu_{V\infty}^2 x}\right)^{\frac{1}{4}} \int_\delta^y \frac{\rho_V}{\rho_{V\infty}} dy, \tag{9.2-31}$$

$$F_{GV} = \frac{\Psi_V}{2\sqrt{2}\left(g\nu_{V\infty}^2 x^3\right)^{\frac{1}{4}}}. \tag{9.2-32}$$

The similarity transformation yields

$$\left(\frac{\rho_V \mu_V}{\rho_{V\infty}\mu_{V\infty}}F_{GV}''\right)' + 3F_{GV}'' F_{GV} - 2F_{FV}'^2 + \left(1 - \frac{\rho_{V\infty}}{\rho_V}\right) = 0, \tag{9.2-33}$$

$$\left(\frac{\rho_V \lambda_V}{\rho_{V\infty}\lambda_{V\infty}}\Theta_{GV}'\right)' + 3\Pr_{V\infty}\frac{c_{pV}}{c_{pV\infty}}F_{GV}\Theta_{GV}' = 0, \tag{9.2-34}$$

$$\left(\frac{\rho_V^2 D}{\rho_{V\infty}^2 D_\infty}\Phi_{GV}'\right)' + 3Sc_\infty F_{GV}\Phi_{GV}' = 0, \tag{9.2-35}$$

$$F_{GL}''' + 3F_{GL}'' F_{GL} - 2F_{GL}'^2 + 1 = 0, \tag{9.2-36}$$

$$\Theta_{GL}'' + 3\Pr_L F_{GL}\Theta_{GL}' = 0. \tag{9.2-37}$$

The boundary and compatibility conditions have the same form as those for the previously considered forced-convection condensation case except that

as $\eta_{GL} \to \infty$:

$$F_{GV\infty}' = 0 ; \tag{9.2-38}$$

at $\eta_{GVi}=0$:

$$F_{GVi} = R_\infty F_{GLi} = \frac{1}{3} R_\infty \dot{M}_{GL} \, . \qquad (9.2\text{-}39)$$

The heat flux q_{wx} and condensation mass flux $\dot{m}_x$, which are obtained from the numerical solution, are expressed as

$$q_{wx}x^{\frac{1}{4}} \;=\; \lambda_L \left(\frac{g}{4\nu_L{}^2}\right)^{\frac{1}{4}} (-\Theta'_{GLw})\,(T_i - T_w)\,, \qquad (9.2\text{-}40)$$

$$\dot{m}_x x^{\frac{1}{4}} \;=\; 3g^{\frac{1}{4}}\left(\frac{\rho_L \nu_L}{2}\right)^{\frac{1}{2}} F_{GLi} \;=\; g^{\frac{1}{4}}\left(\frac{\rho_L \mu_L}{2}\right)^{\frac{1}{2}} \dot{M}_{GL}\,. \qquad (9.2\text{-}41)$$

The heat flux $(q_{wx})_e$ and condensation mass flux $(\dot{m}_x)_e$, which are obtained by solving the simultaneous algebraic equations (4.6-1), (4.6-2), (2-29), and (2-28), are expressed as

$$(q_{wx})_e x^{\frac{1}{4}} \;=\; \lambda_L \left(\frac{g}{4\nu_L{}^2}\right)^{\frac{1}{4}} (\dot{M}_{GL})_e^{-\frac{1}{3}}\,(T_i - T_w)\,, \qquad (9.2\text{-}42)$$

$$(\dot{m}_x)_e\, x^{\frac{1}{4}} \;=\; g^{\frac{1}{4}}\left(\frac{\rho_L \mu_L}{2}\right)^{\frac{1}{2}} \left(\dot{M}_{GL}\right)_e\,, \qquad (9.2\text{-}43)$$

respectively. The comparisons between Eqs. (9.2-40) and (9.2-42) and Eqs. (9.2-41) and (9.2-43) yield the representative concentration W_{1Vr} and temperature T_{Vr} for the representative physical properties for use in the algebraic solution.

Table 9.2-2 shows some examples of numerical results for air-water and ethanol-water mixtures. Figures 9.2-2(a), (b), and (c) show the comparison between the numerical results and corresponding algebraic solutions for AW-1Sa, AW-1Su, and EW-1Sa in Table 9.2-2, respectively. The method of comparison and symbols in the figures are the same as those in Figs. 9.2-1(a), (b), and (c). The conclusion of choosing W_{1Vr} and T_{Vr} is also the same as that in forced-convection condensation, except that a somewhat larger error sometimes appears when the representative physical properties at $(W_{1V\infty} + W_{1Vi})/2$ and $(T_{V\infty} + T_i)/2$ are used.

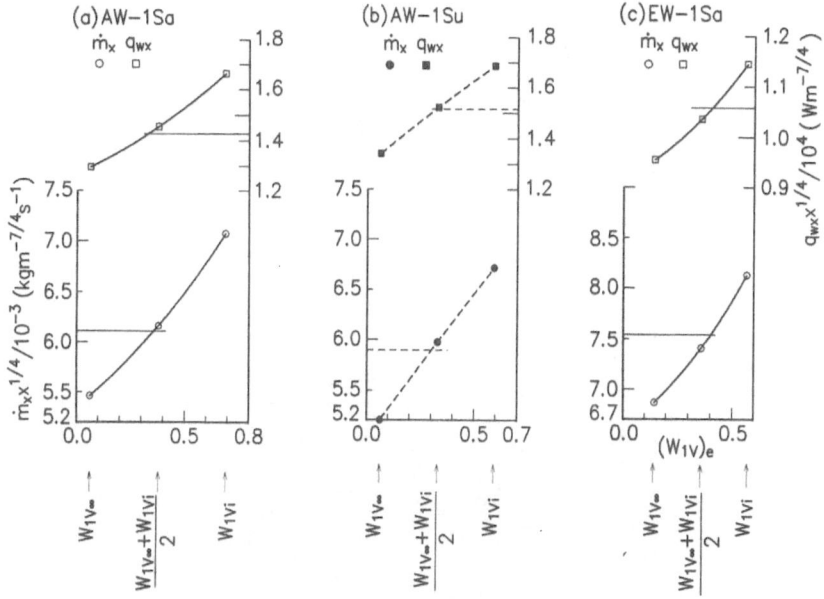

FIGURE 9.2-2. Comparison of algebraic solutions $(q_{wx})_e x^{1/4}$ from Eq. (9.2-42) and $(\dot{m}_x)_e x^{1/4}$ from Eq. (9.2-43) with corresponding numerical solutions $q_{wx} x^{1/4}$ from Eq. (9.2-40) and $\dot{m}_x x^{1/4}$ from Eq. (9.2-41) in the case of free-convection condensation of binary vapor mixtures. Symbols used are the same as in Fig. 9.2-1. (a) AW-1Sa, (b) AW-1Su, (c) EW-1Sa from Table 9.2-2.

TABLE 9.2-2. Numerical results for free-convection condensation of binary vapor mixtures ($p=0.1$ MPa) in the case where the physical properties in the vapor boundary layer depend on concentration and temperature. AW: air-water; EW: ethanol-water; Sa: saturated; Su: superheated.

No.	$\dfrac{T_{V\infty}}{°C}$	$W_{1V\infty}$	η_{GLi}	$\dfrac{F''_{GLw}}{10^{-3}}$	F_{GVi}	$\dfrac{F'_{GVi}}{10^{-3}}$	F''_{GVi}	$-\Phi'_{Gi}$	$-\Theta'_{GLw}$	W_{1Vi}	$\dfrac{T_i}{°C}$	$\dfrac{T_w}{°C}$	$\dfrac{q_{wx}x^{1/4}}{Wm^{-7/4}\times10^3}$	$\dfrac{\dot m_x x^{1/4}}{kgm^{-7/4}s^{-1}\times10^{-3}}$
AW-1Sa	98.5	0.0625	0.2	2.0108	5.9710	2.022	0.13284	0.5837	5.002	0.6985	76.32	74.18	14.306	6.123
AW-2Sa	98.5	0.0625	0.16	1.6058	2.7157	1.2894	0.09678	0.4478	6.251	0.2924	93.37	92.69	6.415	2.785
EW-1Sa	98.0	0.1451	0.22	2.2070	7.3078	2.436	0.11757	0.9219	4.548	0.5748	89.44	86.24	16.128	7.620
AW-1Su	198.5	0.0625	0.2	2.0149	5.6367	2.030	0.2001	0.6458	5.002	0.6020	82.10	79.99	14.784	5.908
AW-2Su	198.5	0.0625	0.16	1.6124	2.6518	1.2999	0.2157	0.5111	6.251	0.2362	94.79	94.05	7.035	2.779
AW-3Su	198.5	0.0625	0.2049	2.0640	6.2944	2.1302	0.1932	0.6471	4.883	0.6950	76.71	74.18	16.515	6.597
AW-4Su	198.5	0.0625	0.1684	1.6965	3.1141	1.4391	0.2151	0.5451	5.940	0.2834	93.60	92.69	8.228	3.264

REFERENCES

1. Fujii, T., Theoretical Consideration of Representative Physical Properties of Liquid Film for Condensation of Saturated Vapours (in preparation).

2. Fujii, T., M. Watabe, and J. B. Lee, Numerical Analysis of Representative Physical Properties of Liquid Film for Forced Convection Condensation of Saturated Vapours (in preparation).

3. Sparrow, E. M., W. J. Minkowycz, and M. Saddy, Forced Convection Condensation in the Presence of Non-condensables and Interfacial Resistance, *Int. J. Heat Mass Transfer*, 10, 1829–1845 (1967).

4. Denny, V. E. and A. F. Mills, Nonsimilar Solutions for Laminar Film Condensation on a Vertical Surface, *Int. J. Heat Mass Transfer*, 12, 965–979 (1969).

5. Shekriladze, I. G. and V. I. Gomelauri, Theoretical Study of Laminar Film Condensation of a Flowing Vapor, *Int. J. Heat Mass Transfer*, 9, 581–591 (1966).

6. McAdams, W. H., "Heat Transmission," p.330, 3rd Ed., McGraw-Hill, New York (1954).

7. Minkowycz, W. J. and E. M. Sparrow, Condensation Heat Transfer in the Presence of Noncondensable, Interfacial Resistance, Super Heating, Variable Properties, and Diffusion, *Int. J. Heat Mass Transfer*, 9, 1125–1144 (1966).

8. Poots, G. and R. G. Miles, Effects of Variable Physical Properties on Laminar Film Condensation of Saturated Steam on a Vertical Flat Plate, *Int. J. Heat Mass Transfer*, 10, 1677–1692 (1967).

9. Voskresenskiy, K. D. , Calculation of heat transfer in film condensation allowing for the temperature dependence of the physical properties of the condensate, *USSR Acad. Sci. OTK* (1948).

10. Labuntsov, D. A. , Effect of temperature dependence of physical parameters of condensate on heat transfer in film condensation of steam, *Teploenergetika*, 4(2), 49–51 (1957).

11. Fujii, T., M. Watabe, and J. B. Lee, Representative Physical Properties of Vapour Boundary Layer for Condensation of Binary Vapours (I. Forced Convection Condensation) (in preparation).

12. Fujii, T., M. Watabe, and J. B. Lee, Representative Physical Properties of Vapour Boundary Layer for Condensation of Binary Vapours (II. Free Convection Condensation) (in preparation).

10

Condensation of Pure Vapors in the Subcritical Region

In the subcritical region, where pressure is a little lower than the critical pressure, all relevant physical properties remarkably depend on temperature, R approaches unity, Pr_L becomes large, and Δh_V becomes small. In the case of free-convection condensation, the buoyancy force acting on the condensate film becomes small as ρ_V approaches ρ_L. The effects of these factors are clarified in this chapter. As an example of the numerical analysis, we use saturated water and saturated carbon dioxide, because their physical properties in the subcritical region are accurately known.

10.1 Forced-Convection Condensation[1]

10.1.1 BASIC EQUATIONS AND SIMILARITY TRANSFORMATION

The basic equations in the case where the physical properties in the condensate film vary with temperature are written as

$$\frac{\partial(\rho_L U_L)}{\partial x} + \frac{\partial(\rho_L V_L)}{\partial y} = 0, \tag{10.1-1}$$

$$\rho_L\left(U_L\frac{\partial U_L}{\partial x} + V_L\frac{\partial U_L}{\partial y}\right) = \frac{\partial}{\partial y}\left(\mu_L\frac{\partial U_L}{\partial y}\right), \tag{10.1-2}$$

$$\rho_L c_{pL}\left(U_L\frac{\partial T_L}{\partial x} + V_L\frac{\partial T_L}{\partial y}\right) = \frac{\partial}{\partial y}\left(\lambda_L\frac{\partial T_L}{\partial y}\right), \tag{10.1-3}$$

$$\frac{\partial U_V}{\partial x} + \frac{\partial V_V}{\partial y} = 0, \tag{10.1-4}$$

$$U_V\frac{\partial U_V}{\partial x} + V_V\frac{\partial U_V}{\partial y} = \nu_V\frac{\partial^2 U_V}{\partial y^2}. \tag{10.1-5}$$

The boundary and compatibility conditions are the same as those in Chapter 2, i.e., Eqs. (2-14)–(2-18) and Eqs. (2-20)–(2-25).

We introduce the following similarity variables and functions for the condensate film:

$$U_L = \frac{\rho_{Ls}}{\rho_L}\frac{\partial \Psi_L}{\partial y}, \qquad\qquad V_L = -\frac{\rho_{Ls}}{\rho_L}\frac{\partial \Psi_L}{\partial x}, \qquad (10.1\text{-}6a,b)$$

$$\eta_{FL} = \left(\frac{U_{V\infty}}{\nu_{Ls}x}\right)^{\frac{1}{2}}\int_0^y \frac{\rho_L}{\rho_{Ls}}dy, \qquad (10.1\text{-}7)$$

$$\Psi_L = (\nu_{Ls}U_{V\infty}x)^{\frac{1}{2}} F_{FL}, \qquad (10.1\text{-}8)$$

$$\Theta_{FL} = \frac{T_s - T_L}{T_s - T_w} \qquad (10.1\text{-}9)$$

and those for the vapor boundary layer are the same as those in Chapter 3, i.e., Eqs. (3.1-2a), (3.1-2b), (3.1-4), and (3.1-6).

The transformation of the basic equations by using these equations yields the following ordinary differential equations:

$$\left(\frac{\rho_L\nu_L}{\rho_{Ls}\nu_{Ls}}F''_{FL}\right)' + \frac{1}{2}F_{FL}F''_{FL} = 0, \qquad (10.1\text{-}10)$$

$$\left(\frac{\rho_L\lambda_L}{\rho_{Ls}\lambda_{Ls}}\Theta'_{FL}\right)' + \frac{1}{2}Pr_{Ls}\frac{c_{pL}}{c_{pLs}}F_{FL}\Theta'_{FL} = 0, \qquad (10.1\text{-}11)$$

$$F'''_{FV} + \frac{1}{2}F_{FV}F''_{FV} = 0 \qquad (10.1\text{-}12)$$

with the boundary conditions

at $\eta_{FL} = 0$:
$$F_{FLw} = 0, \qquad (10.1\text{-}13)$$
$$F'_{FLw} = 0, \qquad (10.1\text{-}14)$$
$$\Theta_{FLw} = 1; \qquad (10.1\text{-}15)$$

as $\eta_{FV} \to \infty$:
$$F'_{FV\infty} = 1; \qquad (10.1\text{-}16)$$

and with the compatibility conditions

at $\eta_{FL} = \eta_{FLi}$ or $\eta_{FV} = 0$:

$$F_{FVi} = \left(\frac{\rho_L\mu_L}{\rho_V\mu_V}\right)^{\frac{1}{2}}_s F_{FLi}, \qquad (10.1\text{-}17)$$

$$F'_{FVi} = F'_{FLi}, \qquad (10.1\text{-}18)$$

$$F''_{FVi} = \left(\frac{\rho_L\mu_L}{\rho_V\mu_V}\right)^{\frac{1}{2}}_s F''_{FLi}, \qquad (10.1\text{-}19)$$

$$\Theta_{FLi} = 0, \qquad (10.1\text{-}20)$$

$$-\Theta'_{FLi} = \frac{\mu_{Ls}\Delta h_V}{\lambda_{Ls}(T_s - T_w)}\dot{M}_{FLs}, \qquad (10.1\text{-}21)$$

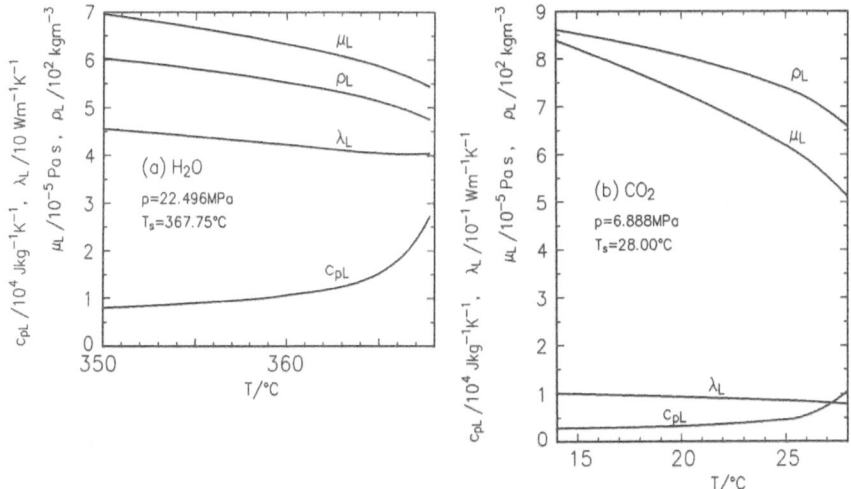

FIGURE 10.1-1. Variation of physical properties with temperature. (a) Water at $p=20.496$ MPa ($T_s = 367.75°C$), (b) carbon dioxide at $p=6.888$ MPa ($T_s = 28.00°C$).

where

$$\dot{M}_{FLs} = \frac{\dot{m}_x x}{\mu_{Ls}(Re_{Lsx})^{\frac{1}{2}}} = \frac{1}{2}F_{FLi}, \tag{10.1-22}$$

$$Re_{Lsx} = \frac{U_{V\infty}x}{\nu_{Ls}}. \tag{10.1-23}$$

10.1.2   EXAMPLES OF NUMERICAL SOLUTIONS

The numerical calculation was made by means of the Runge-Kuta-Verner method, where relevant physical property values were taken from the numerical data in PROPATH[2]. Figures 10.1-1(a) and (b) show the values of ρ_L, μ_L, λ_L, and c_{pL} of water at $p = 20.496$ MPa ($T_s = 367.75°C$) and carbon dioxide at $p = 6.888$ MPa ($T_s = 28.00°C$), respectively. We can see that the variation of c_{pL} in both figures is marked and the variation of μ_L for carbon dioxide is also considerable.

Tables 10.1-1(a) and (b) show a few examples of the numerical results for water and carbon dioxide, respectively. Note that the similarity of these solutions is valid only for given values of T_s and T_w. When the accuracy of these results was checked in the relation of energy balance

$$\frac{d}{dx}\int_0^\delta \rho_L U_L \left(\int_{T_L}^{T_s} c_{pL}dT_L\right)dy = \left(\lambda_L\frac{\partial T_L}{\partial y}\right)_w - \left(\lambda_L\frac{\partial T_L}{\partial y}\right)_s, \tag{10.1-24}$$

the difference between the values of left-hand side and right-hand side was

TABLE 10.1-1. Boundary values, physical properties, and dimensionless numbers in the case of forced-convection condensation of a pure vapor in the subcritical region.

(a) Water
(WA-1 ~5)

$p_c = 22.120$MPa, $T_c = 374.15°C$ $\rho_{Ls} = 554.3$kg/m³, $\mu_{Ls} = 63.2\mu$Pas, $\lambda_{Ls} = 0.424$W/mK, $c_{pLs} = 11.97$kJ/kgK

$p_s = 17.513$MPa, $T_s = 354.7°C$ $\rho_{Vs} = 126.3$kg/m³, $\mu_{Vs} = 24.5\mu$Pas, $\lambda_{Vs} = 0.151$W/mK, $c_{pVs} = 19.47$kJ/kgK

$\Delta h_V = 818.95$kJ/kg

$Pr_{Ls} = 1.78$, $Pr_{Vs} = 3.16$, $R_s = 3.37$

No.	η_{FLi}	F''_{FLw}	F_{FLi}	F'_{FLi}	F''_{FLi}	$-\Theta'_{FLi}$	$-\Theta'_{FLw}$	T_w °C	ρ_{Lw} kg/m³	μ_{Lw} μPas	λ_{Lw} W/mK	c_{pLw} kJ/kgK	Pr_{Lw}	$\dfrac{T_i}{T_w}$	$\dfrac{q_{iz}}{q_{wz}}$	$\left(\dfrac{Nu_{Lwz}}{Re_{Lz}^{1/2}}\right)_r$	$\left(\dfrac{RH}{Pr_L}\right)_r$	$H_r (\tau_T=0.8)$
WA-1	0.5	0.1016	0.0127	0.0511	0.1025	2.006	1.993	354.31	557.1	63.6	0.426	11.69	1.74	0.999	0.998	2.005	0.0107	0.0056
WA-2	0.7	0.1037	0.0256	0.0735	0.1063	1.440	1.414	353.61	561.8	64.2	0.429	11.20	1.68	0.997	0.995	1.438	0.0302	0.0157
WA-3	1.0	0.1067	0.0546	0.1105	0.1142	1.022	0.973	351.44	575.1	65.8	0.437	9.96	1.50	0.991	0.985	1.019	0.0915	0.0457
WA-4	1.5	0.1112	0.1322	0.1817	0.1333	0.712	0.619	343.38	612.9	70.7	0.463	7.86	1.20	0.970	0.954	0.705	0.3310	0.1449
WA-5	2.0	0.1146	0.2510	0.2647	0.1579	0.569	0.440	327.80	664.8	78.1	0.506	6.40	0.99	0.931	0.905	0.558	0.8364	0.2988

(WB-1 ~5)

$p_c = 22.120$MPa, $T_c = 374.15°C$ $\rho_{Ls} = 520.8$kg/m³, $\mu_{Ls} = 59.4\mu$Pas, $\lambda_{Ls} = 0.409$W/mK, $c_{pLs} = 15.42$kJ/kgK

$p_s = 18.947$MPa, $T_s = 361.2°C$ $\rho_{Vs} = 148.8$kg/m³, $\mu_{Vs} = 25.9\mu$Pas, $\lambda_{Vs} = 0.185$W/mK, $c_{pVs} = 27.06$kJ/kgK

$\Delta h_V = 696.81$kJ/kg

$Pr_{Ls} = 2.24$, $Pr_{Vs} = 3.79$, $R_s = 2.83$

No.	η_{FLi}	F''_{FLw}	F_{FLi}	F'_{FLi}	F''_{FLi}	$-\Theta'_{FLi}$	$-\Theta'_{FLw}$	T_w °C	ρ_{Lw} kg/m³	μ_{Lw} μPas	λ_{Lw} W/mK	c_{pLw} kJ/kgK	Pr_{Lw}	$\dfrac{T_i}{T_w}$	$\dfrac{q_{iz}}{q_{wz}}$	$\left(\dfrac{Nu_{Lwz}}{Re_{Lz}^{1/2}}\right)_r$	$\left(\dfrac{RH}{Pr_L}\right)_r$	$H_r (\tau_T=0.8)$
WB-1	0.5	0.1199	0.0151	0.0604	0.1215	2.006	1.991	360.82	524.6	59.8	0.411	14.78	2.15	0.999	0.997	2.008	0.0107	0.0083
WB-2	0.7	0.1212	0.0301	0.0864	0.1256	1.441	1.412	360.15	530.9	60.5	0.413	13.80	2.02	0.997	0.992	1.443	0.0297	0.0228
WB-2	1.0	0.1220	0.0629	0.1277	0.1339	1.024	0.969	358.10	547.7	62.5	0.420	11.76	1.75	0.990	0.979	1.028	0.0885	0.0641
WB-3	1.5	0.1221	0.1463	0.2024	0.1525	0.714	0.616	350.85	589.3	67.7	0.446	8.86	1.34	0.967	0.940	0.717	0.3074	0.1878
WB-4	2.0	0.1219	0.2679	0.2840	0.1751	0.568	0.439	337.37	641.1	74.6	0.486	6.89	1.06	0.928	0.886	0.570	0.7523	0.3680

TABLE 10.1-1. (continued)

(WC-1~5)*

$p_c = 22.120$MPa,	$T_c = 374.15$°C	$\rho_{Ls} = 475.4$kg/m³,	$\mu_{Ls} = 54.3\mu$Pas,	$\lambda_{Ls} = 0.404$W/mK,	$c_{pLs} = 27.37$kJ/kgK
$p_s = 20.496$MPa,	$T_s = 367.75$°C	$\rho_{Vs} = 182.9$kg/m³,	$\mu_{Vs} = 28.1\mu$Pas,	$\lambda_{Vs} = 0.255$W/mK,	$c_{pVs} = 50.73$kJ/kgK
		$\Delta h_V = 533.01$kJ/kg			
		$Pr_{Ls} = 3.68$,	$Pr_{Vs} = 5.58$,	$R_s = 2.24$	

No.	η_{FLi}	F''_{FLw}	F_{FLi}	F'_{FLi}	F''_{FLi}	$-\Theta'_{FLi}$	$-\Theta'_{FLw}$	$\frac{T}{°C}$	$\frac{\rho_{Lw}}{kg/m^3}$	$\frac{\mu_{Lw}}{\mu Pas}$	$\frac{\lambda_{Lw}}{W/mK}$	$\frac{c_{pLw}}{kJ/kgK}$	Pr_{Lw}	$\frac{T_i}{T_w}$	$\frac{q_{ix}}{q_{wx}}$	$\left(\frac{Nu_{Lwx}}{Re_{Lx}^{1/2}}\right)_r$	$\left(\frac{RH}{Pr_L}\right)_r$	$\frac{H_r}{(\tau_r=0.8)}$
WC-1	0.5	0.1496	0.0188	0.0756	0.1529	2.002	1.993	367.41	481.1	55.0	0.403	24.84	3.39	0.998	0.994	2.015	0.0106	0.0170
WC-2	0.7	0.1485	0.0371	0.1070	0.1572	1.432	1.414	366.82	490.2	56.0	0.403	21.35	2.97	0.996	0.985	1.457	0.0290	0.0451
WC-3	1.0	0.1442	0.0752	0.1539	0.1655	1.010	0.972	365.08	512.5	58.5	0.405	15.43	2.23	0.988	0.963	1.051	0.0834	0.1180
WC-4	1.5	0.1371	0.1659	0.2318	0.1826	0.693	0.617	359.17	558.1	63.9	0.426	10.27	1.54	0.964	0.910	0.743	0.2756	0.2906
WC-5	2.0	0.1318	0.2909	0.3109	0.2020	0.544	0.441	348.58	608.7	70.3	0.460	7.78	1.19	0.925	0.846	0.592	0.6520	0.4836

TABLE 10.1-1. (continued)

(b) Carbon dioxide

(CA-1~4)

$p_c = 7.383$ MPa, $\quad T_c = 31.06°C$
$p_s = 5.9863$ MPa, $\quad T_s = 21.9°C$

$\rho_{Ls} = 751.8$ kg/m³, $\quad \mu_{Ls} = 63.9\,\mu$Pas, $\quad \lambda_{Ls} = 0.088$ W/mK, $\quad c_{pLs} = 4.78$ kJ/kgK
$\rho_{Vs} = 209.9$ kg/m³, $\quad \mu_{Vs} = 19.2\,\mu$Pas, $\quad \lambda_{Vs} = 0.032$ W/mK, $\quad c_{pVs} = 5.27$ kJ/kgK
$\Delta h_V = 141.18$ kJ/kg
$Pr_{Ls} = 3.48$, $\quad Pr_{Vs} = 3.15$, $\quad R_s = 3.46$

No.	η_{FLi}	F''_{FLw}	F_{FLi}	F'_{FLi}	F''_{FLi}	$-\Theta'_{FLi}$	$-\Theta'_{FLw}$	$\frac{T}{°C}$	$\frac{\rho_{Lw}}{kg/m^3}$	$\frac{\mu_{Lw}}{\mu Pas}$	$\frac{\lambda_{Lw}}{W/mK}$	$\frac{c_{pLw}}{kJ/kgK}$	Pr_{Lw}	$\frac{T_i}{T_w}$	$\frac{q_{ir}}{q_{wr}}$	$\left(\frac{Nu_{Lwz}}{Re_{Lz}^{1/2}}\right)_r$	$\left(\frac{RH}{Pr_L}\right)_r$	H_r $(\tau_T=0.8)$
CA-1	0.8	0.0980	0.0322	0.0814	0.1057	1.270	1.223	20.60	773.0	67.3	0.090	4.15	3.11	0.996	0.986	1.277	0.0441	0.0427
CA-2	1.0	0.0967	0.0505	0.1033	0.1107	1.030	0.962	19.38	790.2	70.2	0.092	3.77	2.89	0.992	0.974	1.041	0.0861	0.0804
CA-3	1.5	0.0905	0.1122	0.1584	0.1268	0.721	0.609	13.89	848.9	81.3	0.099	2.96	2.43	0.976	0.928	0.743	0.2812	0.2295
CA-4	2.0	0.0821	0.1916	0.2111	0.1459	0.574	0.431	4.73	918.2	97.7	0.110	2.48	2.21	0.952	0.870	0.607	0.6269	0.4341

(CB-1~4)

$p_c = 7.383$ MPa, $\quad T_c = 31.06°C$
$p_s = 6.4311$ MPa, $\quad T_s = 25.0°C$

$\rho_{Ls} = 710.6$ kg/m³, $\quad \mu_{Ls} = 57.9\,\mu$Pas, $\quad \lambda_{Ls} = 0.084$ W/mK, $\quad c_{pLs} = 6.35$ kJ/kgK
$\rho_{Vs} = 242.5$ kg/m³, $\quad \mu_{Vs} = 20.5\,\mu$Pas, $\quad \lambda_{Vs} = 0.036$ W/mK, $\quad c_{pVs} = 7.77$ kJ/kgK
$\Delta h_V = 119.76$ kJ/kg
$Pr_{Ls} = 4.40$, $\quad Pr_{Vs} = 4.46$, $\quad R_s = 2.88$

No.	η_{FLi}	F''_{FLw}	F_{FLi}	F'_{FLi}	F''_{FLi}	$-\Theta'_{FLi}$	$-\Theta'_{FLw}$	$\frac{T}{°C}$	$\frac{\rho_{Lw}}{kg/m^3}$	$\frac{\mu_{Lw}}{\mu Pas}$	$\frac{\lambda_{Lw}}{W/mK}$	$\frac{c_{pLw}}{kJ/kgK}$	Pr_{Lw}	$\frac{T_i}{T_w}$	$\frac{q_{ir}}{q_{wr}}$	$\left(\frac{Nu_{Lwz}}{Re_{Lz}^{1/2}}\right)_r$	$\left(\frac{RH}{Pr_L}\right)_r$	H_r $(\tau_T=0.8)$
CB-1	0.8	0.1141	0.0377	0.0957	0.1259	1.276	1.216	23.78	738.1	61.8	0.086	5.01	3.60	0.995	0.980	1.287	0.0430	0.0613
CB-2	1.0	0.1109	0.0583	0.1200	0.1312	1.036	0.956	22.66	757.7	64.8	0.088	4.38	3.22	0.991	0.964	1.053	0.0827	0.1115
CB-3	1.5	0.1007	0.1256	0.1784	0.1473	0.724	0.607	17.80	819.3	75.4	0.095	3.24	2.57	0.974	0.910	0.757	0.2625	0.2922
CB-4	2.0	0.0900	0.2099	0.2322	0.1657	0.573	0.433	9.79	887.8	90.0	0.105	2.64	2.26	0.949	0.845	0.621	0.5759	0.5221

TABLE 10.1-1. (continued)

(CC-1 ~5)

| $p_c = 7.383$MPa, | $T_c = 31.06$°C |
| $p_s = 6.8876$MPa, | $T_s = 28.0$°C |

$\rho_{Ls} = 658.0$kg/m³, $\mu_{Ls} = 51.2\mu$Pas, $\lambda_{Ls} = 0.079$W/mK, $c_{pLs} = 10.48$kJ/kgK

$\rho_{Vs} = 288.8$kg/m³, $\mu_{Vs} = 22.5\mu$Pas, $\lambda_{Vs} = 0.041$W/mK, $c_{pVs} = 14.87$kJ/kgK

$\Delta h_V = 91.52$kJ/kg

$Pr_{Ls} = 6.81$, $Pr_{Vs} = 8.17$, $R_s = 2.27$

No.	η_{FLi}	F''_{FLw}	F_{FLi}	F'_{FLi}	F''_{FLi}	$-\Theta'_{FLi}$	$-\Theta'_{FLw}$	$\frac{T_w}{°C}$	$\frac{\rho_{Lw}}{kg/m^3}$	$\frac{\mu_{Lw}}{\mu Pas}$	$\frac{\lambda_{Lw}}{W/mK}$	$\frac{c_{pLw}}{kJ/kgK}$	Pr_{Lw}	$\frac{T_i}{T_w}$	$\frac{q_{iz}}{q_{wx}}$	$\left(\frac{Nu_{Lwz}}{Re_{Lz}^{1/2}}\right)_r$	$\left(\frac{RH}{Pr_L}\right)_r$	$\frac{H_L}{(\tau_T=0.8)}$
CC-1	0.7	0.1400	0.0354	0.1027	0.1549	1.4550	1.3906	27.28	685.3	54.6	0.081	7.42	4.99	0.996	0.976	1.473	0.0280	0.0760
CC-2	0.8	0.1371	0.0457	0.1170	0.1573	1.2819	1.2064	26.94	695.4	55.9	0.082	6.68	4.55	0.994	0.966	1.305	0.0413	0.1074
CC-3	1.0	0.1308	0.0695	0.1441	0.1626	1.0410	0.9488	26.02	717.8	58.9	0.084	5.50	3.86	0.989	0.943	1.074	0.0781	0.1835
CC-4	1.5	0.1148	0.1440	0.2062	0.1781	0.7220	0.6060	22.07	781.7	68.7	0.091	3.70	2.81	0.972	0.872	0.779	0.2409	0.4188
CC-5	2.0	0.1010	0.2349	0.2610	0.1952	0.5635	0.4363	15.61	848.0	81.1	0.099	2.91	2.39	0.946	0.799	0.641	0.5221	0.6896

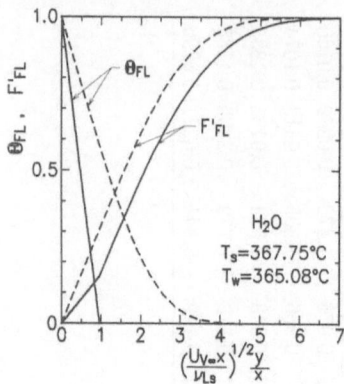

FIGURE 10.1-2. An example of the distributions of F'_{FL}, F'_{FV}, and Θ_{FL}. Solid lines are for condensation and broken lines are for single phase convection of water at $p=20.496$ MPa.

within 10^{-3} percent. We can see that $\lambda_{Ls}(-\Theta'_{FLi})$ which is proportional to q_{ix} is smaller than $(\lambda_{LW}\rho_{Lw}/\rho_{Ls})(-\Theta'_{FLw})$ which is proportional to q_{wx}, and $\mu_{Ls}F''_{FLi}$ which is proportional to τ_i is smaller than $(\mu_{Lw}\rho_{Lw}/\rho_{Ls})F''_{FLw}$ which is proportional to τ_w, and the differences are marked as $(T_s - T_w)$ is large and when T_s is near the critical temperature T_c. The effect of the convection term upon the temperature distribution and the effect of the inertia term upon the velocity distribution are marked in the subcritical region.

Figure 10.1-2 shows an example of the distributions of F'_{FL}, F'_{FV}, and Θ_{FL} with solid lines, which are propotional to the velocity component U_L and U_V and temperature T_L, respectively, where a dimensionless abscissa represents the same enlargement rate in the condensate film and in the vapor boundary layer as

$$\int_0^{\eta_{FL}} \left(\frac{\rho_{Ls}}{\rho_L}\right) d\eta_{FL} = \left(\frac{U_{V\infty}x}{\nu_{Ls}}\right)^{\frac{1}{2}} \frac{y}{x} \qquad (10.1\text{-}25\text{a})$$

and

$$\left(\frac{\nu_{Vs}}{\nu_{Ls}}\right)^{\frac{1}{2}} \eta_{FV} = \left(\frac{U_{V\infty}x}{\nu_{Ls}}\right)^{\frac{1}{2}} \frac{(y-\delta)}{x} . \qquad (10.1\text{-}25\text{b})$$

The broken lines in the figure show the distributions of velocity $(F'_{FV})_s$ and temperature $(\Theta_{FL})_s$ in the single phase forced convection of water with the same conditions in temperatures $T_{V\infty}(=T_s)$ and T_w and pressure. As for the velocity distributions, there is no large difference between the single phase and the two-phase case, while as for the temperature distributions there is still a somewhat large difference between them.

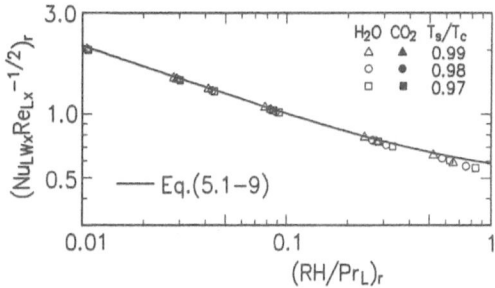

FIGURE 10.1-3. Correlation between $(Nu_{Lwx}/Re_{Lx}^{1/2})_r$ and $(RH/Pr_L)_r$.

10.1.3   WALL HEAT FLUX AND CONDENSATION MASS FLUX

The heat transfer coefficient α_{wx} and the corresponding Nusselt number Nu_{Lwx} are defined, respectively, as

$$\alpha_{wx} = \frac{q_{wx}}{T_s - T_w} = (-\Theta'_{FLw})\frac{\lambda_{Lw}\rho_{Lw}}{\rho_{Ls}}\left(\frac{U_{V\infty}}{\nu_{Ls}x}\right)^{\frac{1}{2}}, \quad (10.1\text{-}26)$$

$$Nu_{Lwx} = \frac{\alpha_{wx}x}{\lambda_{Lw}} = (-\Theta'_{FLw})\left(\frac{\rho_{Lw}}{\rho_{Ls}}\right)Re_{Lsx}^{\frac{1}{2}}. \quad (10.1\text{-}27)$$

Figure 10.1-3 shows the relation between $(Nu_{Lwx}/Re_{Lx}^{1/2})_r$ and $(RH/Pr_L)_r$, where the former value is calculated from

$$\left(\frac{Nu_{Lwx}}{Re_{Lx}^{\frac{1}{2}}}\right)_r = (-\Theta'_{FLw})\left(\frac{\rho_{Lw}\lambda_{Lw}}{\rho_{Ls}\lambda_{Lr}}\right)\left(\frac{\nu_{Lr}}{\nu_{Ls}}\right)^{\frac{1}{2}} \quad (10.1\text{-}28)$$

and the representative physical properties for both values are evaluated at $r_T = 1/3$ after the conclusion in Chapter 9. The solid line in the figure represents Eq. (5.1-9). We can see that the data can be correlated well by the equation, although the formers become a little smaller than the latter with the increase of $(RH/Pr_L)_r$.

Figure 10.1-4 shows the relation between q_{ix}/q_{wx} and H_r, which is evaluated at $r_T = 0.8$. These data can be correlated fairly well by Eq. (5.1-12), as seen in the figure.

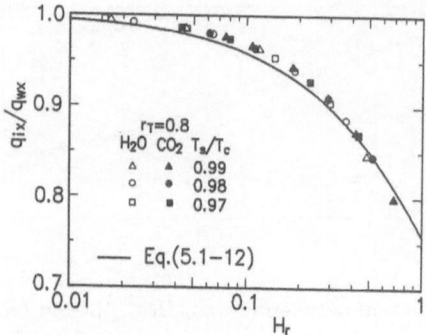

FIGURE 10.1-4. Correlation between q_{ix}/q_{wx} and H_r for forced-convection condensation.

10.2 Free-Convection Condensation[3]

10.2.1 BASIC EQUATIONS AND SIMILARITY TRANSFORMATION

The equation of momentum conservation in the condensate film for free-convection condensation is written as

$$\rho_L \left(U_L \frac{\partial U_L}{\partial x} + V_L \frac{\partial U_L}{\partial y} \right) = \frac{\partial}{\partial y} \left(\mu_L \frac{\partial U_L}{\partial y} \right) + g \left(\rho_L - \rho_{Vs} \right), \quad (10.2\text{-}1)$$

the other basic equations for the condensate film are the same as those for the forced-convection condensation in Section 10.1. The similarity variable and dimensionless stream function for the condensate film are defined as

$$\eta_{GL} = c_L x^{-\frac{1}{4}} \int_0^y \frac{\rho_L}{\rho_{Ls}} dy, \quad (10.2\text{-}2)$$

$$\Psi_{GL} = 4 c_L \nu_{Ls} x^{\frac{3}{4}} F_{GL}, \quad (10.2\text{-}3)$$

where

$$c_L = \left[\frac{g \left(1 - \dfrac{\rho_{Vs}}{\rho_{Ls}} \right)}{4 \nu_{Ls}^2} \right]^{\frac{1}{4}}, \quad (10.2\text{-}4)$$

while Eqs. (4.1-2) and (4.1-4) are used for the similarity variable and the dimensionless stream function in the vapor boundary layer, respectively.

The transformed ordinary differential equations are

$$\left(\frac{\rho_L \mu_L}{\rho_{Ls} \mu_{Ls}} F''_{GL}\right)' + 3F_{GL} F''_{GL} - 2F'^2_{GL} + \frac{\left(1 - \frac{\rho_{Vs}}{\rho_L}\right)}{\left(1 - \frac{\rho_{Vs}}{\rho_{Ls}}\right)} = 0, \quad (10.2\text{-}5)$$

$$\left(\frac{\lambda_L \rho_L}{\lambda_{Ls} \rho_{Ls}} \Theta'_{GL}\right)' + 3\mathrm{Pr}_{Ls}\frac{c_{pL}}{c_{pLs}} F_{GL}\Theta'_{GL} = 0, \quad (10.2\text{-}6)$$

$$F'''_{GV} + 3F''_{GV} F_{GV} - 2F'^2_{GV} = 0 \quad (10.2\text{-}7)$$

with the boundary conditions

at $\eta_{GL} = 0$:
$$\begin{aligned} F_{GL} &= 0, & (10.2\text{-}8) \\ F'_{GL} &= 0, & (10.2\text{-}9) \\ \Theta_{GL} &= 1; & (10.2\text{-}10) \end{aligned}$$

as $\eta_{GV} \to \infty$:
$$F'_{GL} = 0; \quad (10.2\text{-}11)$$

and with the compatibility conditions

at $\eta_{GL} = \eta_{GLi}$ or $\eta_{GV} = 0$:

$$F_{GVi} = R_s \left(1 - \frac{\rho_{Vs}}{\rho_{Ls}}\right)^{\frac{1}{4}} F_{GLi}, \quad (10.2\text{-}12)$$

$$F'_{GVi} = \left(1 - \frac{\rho_{Vs}}{\rho_{Ls}}\right)^{\frac{1}{2}} F'_{GLi}, \quad (10.2\text{-}13)$$

$$F''_{GVi} = R_s \left(1 - \frac{\rho_{Vs}}{\rho_{Ls}}\right)^{\frac{3}{4}} F''_{GLi}, \quad (10.2\text{-}14)$$

$$\Theta_{GLi} = 0, \quad (10.2\text{-}15)$$

$$-\Theta'_{GLi} = \frac{\mathrm{Pr}_{Ls}}{H_S} \dot{M}_{GLs}, \quad (10.2\text{-}16)$$

where

$$R_s = \left(\frac{\rho_{Ls} \mu_{Ls}}{\rho_{Vs} \mu_{Vs}}\right)^{\frac{1}{2}}, \quad (10.2\text{-}17)$$

$$\dot{M}_{GLs} = \frac{\dot{m}_x x}{\mu_{Ls} \left(\frac{Ga_s}{4}\right)^{\frac{1}{4}}} = 3F_{GLi}. \quad (10.2\text{-}18)$$

10.2.2   EXAMPLES OF NUMERICAL SOLUTIONS

The method of numerical calculation and specified conditions in this case are almost identical to those in the forced-convection condensation case.

Tables 10.2-1(a) and (b) show some examples of obtained boundary values for water and carbon dioxide, respectively. Figure 10.2-1 shows an example of distributions of velocity and temperature, where the abscissa is taken as

$$c_L x^{-\frac{1}{4}} y = \int_0^{\eta_{GL}} \frac{\rho_{Ls}}{\rho_L} d\eta_{GL} \qquad (10.2\text{-}19)$$

and

$$c_L x^{-\frac{1}{4}} (y - \delta) = \left(\frac{\nu_{Vs}}{\nu_{Ls}}\right)^{\frac{1}{2}} \left(1 - \frac{\rho_{Vs}}{\rho_{Ls}}\right)^{\frac{1}{4}} \eta_{GV} \qquad (10.2\text{-}20)$$

for the condensate film and the vapor boundary layer, respectively, and the dimensionless velocity in the ordinate is taken as

$$\frac{U_L}{4c_L{}^2 \nu_{Ls} x^{\frac{1}{2}}} = F'_{GL} \qquad (10.2\text{-}21)$$

and

$$\frac{U_V}{4c_L{}^2 \nu_{Ls} x^{\frac{1}{2}}} = \left(1 - \frac{\rho_{Vs}}{\rho_{Ls}}\right)^{-\frac{1}{2}} F'_{GV} \qquad (10.2\text{-}22)$$

for the condensate film and the vapor boundary layer, respectively. We can see in the figure that the maximum velocity appears in the condensate film. The other characteristics inherent in the subcritical region which are seen in the boundary values shown in Tables 10.2-1(a) and (b) and in the distributions shown in Fig. 10.2-1 are similar to those in forced-convection condensation.

10.2.3   WALL HEAT FLUX AND CONDENSATION MASS FLUX

The heat transfer coefficient α_{wx} and the corresponding Nusselt number Nu_{Lwx} are defined, respectively, as

$$\alpha_{wx} = \frac{q_{wx}}{(T_s - T_w)} = (-\Theta'_{GLw})\lambda_{Lw} \frac{\rho_{Lw}}{\rho_{Ls}} c_L x^{-\frac{1}{4}} , \qquad (10.2\text{-}23)$$

$$Nu_{Lwx} = \frac{\alpha_{wx} x}{\lambda_{Lw}} = \frac{(-\Theta'_{GLw})\rho_{Lw}}{\rho_{Ls}} \left(\frac{Gr_{Lsx}}{4}\right)^{\frac{1}{4}} , \qquad (10.2\text{-}24)$$

where

$$Gr_{Lsx} = \frac{x^3 g \left(1 - \dfrac{\rho_{Vs}}{\rho_{Ls}}\right)}{\nu_{Ls}{}^2} . \qquad (10.2\text{-}25)$$

TABLE 10.2-1. Boundary values, physical properties, and dimensionless numbers in the case of free-convection condensation of a pure vapor in the subcritical region.

(a) Water

(WA-1 ~4)

$p_c = 22.120$MPa, $T_c = 374.15°C$ $\rho_{Ls} = 554.3$kg/m³, $\mu_{Ls} = 63.2\mu$Pas, $\lambda_{Ls} = 0.424$W/mK, $c_{pLs} = 11.97$kJ/kgK
$p_s = 17.513$MPa, $T_s = 354.7°C$ $\rho_{Vs} = 126.3$kg/m³, $\mu_{Vs} = 24.5\mu$Pas, $\lambda_{Vs} = 0.151$W/mK, $c_{pVs} = 19.47$kJ/kgK
$\Delta h_V = 818.95$kJ/kg
$Pr_{Ls} = 1.78$, $Pr_{Vs} = 3.16$, $R_s = 3.37$

No.	η_{GLi}	F''_{GLw}	F_{GLi}	F'_{GLi}	F''_{GLi}	$-\Theta'_{GLi}$	$-\Theta'_{GLw}$	$\frac{T_w}{°C}$	$\frac{\rho_{Lw}}{kg/m^3}$	$\frac{\mu_{Lw}}{\mu Pas}$	$\frac{\lambda_{Lw}}{W/mK}$	$\frac{c_{pLw}}{kJ/kgK}$	Pr_{Lw}	$\frac{q_{iz}}{q_{wz}}$	$\left(\frac{Nu_{Lwz}}{(Gr/4)^{1/4}}\right)_r$	$\left(\frac{Pr_L}{H}\right)_r$	$\frac{H_r}{(\tau_T=0.9)}$
WA-1	0.3	0.2884	0.0086	0.0429	-0.0047	3.355	3.305	353.76	560.9	64.0	0.428	11.30	1.69	0.995	3.357	129.9072	0.0137
WA-2	0.6	0.4576	0.0549	0.1338	-0.0362	1.774	1.551	343.38	612.9	70.7	0.463	7.86	1.20	0.948	1.783	10.9624	0.1544
WA-3	0.9	0.4932	0.1313	0.2048	-0.0995	1.292	0.961	317.50	691.7	82.3	0.530	5.93	0.92	0.864	1.310	3.3790	0.4428
WA-4	1.0	0.4900	0.1600	0.2206	-0.1246	1.196	0.849	305.71	718.8	87.0	0.553	5.52	0.87	0.833	1.220	2.5757	0.5543

(WB-1 ~4)

$p_c = 22.120$MPa, $T_c = 374.15°C$ $\rho_{Ls} = 520.8$kg/m³, $\mu_{Ls} = 59.4\mu$Pas, $\lambda_{Ls} = 0.409$W/mK, $c_{pLs} = 15.42$kJ/kgK
$p_s = 18.947$MPa, $T_s = 361.2°C$ $\rho_{Vs} = 148.8$kg/m³, $\mu_{Vs} = 25.9\mu$Pas, $\lambda_{Vs} = 0.185$W/mK, $c_{pVs} = 27.06$kJ/kgK
$\Delta h_V = 696.81$kJ/kg
$Pr_{Ls} = 2.24$, $Pr_{Vs} = 3.79$, $R_s = 2.83$

No.	η_{GLi}	F''_{GLw}	F_{GLi}	F'_{GLi}	F''_{GLi}	$-\Theta'_{GLi}$	$-\Theta'_{GLw}$	$\frac{T_w}{°C}$	$\frac{\rho_{Lw}}{kg/m^3}$	$\frac{\mu_{Lw}}{\mu Pas}$	$\frac{\lambda_{Lw}}{W/mK}$	$\frac{c_{pLw}}{kJ/kgK}$	Pr_{Lw}	$\frac{q_{iz}}{q_{wz}}$	$\left(\frac{Nu_{Lwz}}{(Gr/4)^{1/4}}\right)_r$	$\left(\frac{Pr_L}{H}\right)_r$	$\frac{H_r}{(\tau_T=0.9)}$
WB-1	0.4	0.3596	0.0192	0.0709	-0.0126	2.542	2.441	358.92	541.4	61.8	0.4173	12.44	1.84	0.982	2.559	44.7833	0.0492
WB-2	0.5	0.4130	0.0344	0.1013	-0.0237	2.071	1.910	356.17	560.9	64.1	0.4274	10.63	1.60	0.964	2.096	20.5252	0.1053
WB-3	0.8	0.4847	0.1021	0.1816	-0.0789	1.398	1.108	339.05	635.6	73.9	0.4815	7.03	1.08	0.879	1.440	4.7775	0.3974
WB-4	1.0	0.4896	0.1587	0.2179	-0.1285	1.179	0.851	320.40	688.3	81.8	0.5268	5.96	0.92	0.814	1.234	2.6164	0.6512

TABLE 10.2-1. (continued)

(WC-1 ~4)

$p_c = 22.120\text{MPa}$,	$T_c = 374.15°\text{C}$	$\rho_{Ls} = 475.4\text{kg/m}^3$,	$\mu_{Ls} = 54.3\mu\text{Pas}$,	$\lambda_{Ls} = 0.404\text{W/mK}$,	$c_{pLs} = 27.37\text{kJ/kgK}$
$p_s = 20.496\text{MPa}$,	$T_s = 367.75°\text{C}$	$\rho_{Vs} = 182.9\text{kg/m}^3$,	$\mu_{Vs} = 28.1\mu\text{Pas}$,	$\lambda_{Vs} = 0.255\text{W/mK}$,	$c_{pVs} = 50.73\text{kJ/kgK}$
		$\Delta h_V = 533.01\text{kJ/kg}$			
		$Pr_{Ls} = 3.68$,	$Pr_{Vs} = 5.58$,	$R_s = 2.24$	

No.	η_{GLi}	F''_{GLw}	F_{GLi}	F'_{GLi}	F''_{GLi}	$-\Theta'_{GLi}$	$-\Theta'_{GLw}$	T_w °C	ρ_{Lw} kg/m³	μ_{Lw} μPas	λ_{Lw} W/mK	c_{pLw} kJ/kgK	Pr_{Lw}	$\frac{q_{ix}}{q_{wx}}$	$\left(\frac{Nu_{Lwx}}{(GR/4)^{1/4}}\right)_r$	$\left(\frac{Pr_L}{H}\right)_r$	$\frac{H_L}{(\tau_T=0.9)}$
WC-1	0.4	0.3534	0.0188	0.0695	-0.0148	2.508	2.459	366.14	499.8	57.1	0.403	18.42	2.61	0.973	2.574	45.7865	0.0790
WC-2	0.6	0.4399	0.0524	0.1269	-0.0429	1.706	1.567	361.15	545.2	62.4	0.419	11.26	1.68	0.916	1.828	11.7809	0.2825
WC-3	0.8	0.4798	0.1003	0.1771	-0.0850	1.329	1.121	351.52	596.5	68.7	0.451	8.27	1.26	0.846	1.465	4.9200	0.5597
WC-4	1.0	0.4924	0.1578	0.2148	-0.1371	1.109	0.862	337.13	647.8	75.6	0.492	6.66	1.02	0.775	1.256	2.6469	0.8409

(b) Carbon dioxide

(CA-1 ~3)

$p_c = 7.383\text{MPa}$,	$T_c = 31.06°\text{C}$	$\rho_{Ls} = 751.8\text{kg/m}^3$,	$\mu_{Ls} = 63.9\mu\text{Pas}$,	$\lambda_{Ls} = 0.088\text{W/mK}$,	$c_{pLs} = 4.78\text{kJ/kgK}$
$p_s = 5.9863\text{MPa}$,	$T_s = 21.9°\text{C}$	$\rho_{Vs} = 209.9\text{kg/m}^3$,	$\mu_{Vs} = 19.2\mu\text{Pas}$,	$\lambda_{Vs} = 0.032\text{W/mK}$,	$c_{pVs} = 5.27\text{kJ/kgK}$
		$\Delta h_V = 141.18\text{kJ/kg}$			
		$Pr_{Ls} = 3.48$,	$Pr_{Vs} = 3.15$,	$R_s = 3.46$	

No.	η_{GLi}	F''_{GLw}	F_{GLi}	F'_{GLi}	F''_{GLi}	$-\Theta'_{GLi}$	$-\Theta'_{GLw}$	T_w °C	ρ_{Lw} kg/m³	μ_{Lw} μPas	λ_{Lw} W/mK	c_{pLw} kJ/kgK	Pr_{Lw}	$\frac{q_{ix}}{q_{wx}}$	$\left(\frac{Nu_{Lwx}}{(Gr/4)^{1/4}}\right)_r$	$\left(\frac{Pr_L}{H}\right)_r$	$\frac{H_L}{(\tau_T=0.9)}$
CA-1	0.4	0.3444	0.0186	0.0694	-0.0104	2.563	2.416	19.66	786.4	69.5	0.091	3.85	2.93	0.974	2.597	47.0325	0.0737
CA-2	0.6	0.3972	0.0494	0.1225	-0.0305	1.798	1.522	13.42	853.1	82.2	0.100	2.92	2.41	0.916	1.879	13.0623	0.2606
CA-3	0.8	0.3908	0.0883	0.1631	-0.0584	1.432	1.075	2.87	930.2	101.0	0.112	2.42	2.18	0.843	1.562	6.1808	0.5323

TABLE 10.2-1. (continued)

(CB-1 ~3)

$p_c = 7.383$MPa,	$T_c = 31.06$°C	$\rho_{Ls} = 710.6$kg/m³,	$\mu_{Ls} = 57.9\mu$Pas,	$\lambda_{Ls} = 0.084$W/mK,	$c_{PLs} = 6.35$kJ/kgK
$p_s = 6.4311$MPa,	$T_s = 25.0$°C	$\rho_{Vs} = 242.5$kg/m³,	$\mu_{Vs} = 20.5\mu$Pas,	$\lambda_{Vs} = 0.036$W/mK,	$c_{PVs} = 7.77$kJ/kgK

$\Delta h_V = 119.76$kJ/kg

$Pr_{Ls} = 4.40,$ $Pr_{Vs} = 4.46,$ $R_s = 2.88$

No.	η_{GLi}	F''_{GLw}	F_{GLi}	F'_{GLi}	F''_{GLi}	$-\Theta'_{GLi}$	$-\Theta'_{GLw}$	T_w °C	$\frac{\rho_{Lw}}{\text{kg/m}^3}$	$\frac{\mu_{Lw}}{\mu\text{Pas}}$	$\frac{\lambda_{Lw}}{\text{W/mK}}$	$\frac{c_{PLw}}{\text{kJ/kgK}}$	Pr_{Lw}	$\frac{q_{iz}}{q_{wz}}$	$\left(\frac{Nu_{Lwz}}{(Gr/4)^{1/4}}\right)_r$	$\left(\frac{Pr_L}{H}\right)_r$	$\frac{H_L}{(\tau_T=0.9)}$
CB-1	0.4	0.3407	0.0184	0.0684	-0.0117	2.566	2.413	23.22	748.4	63.4	0.087	4.66	3.39	0.968	2.608	47.8214	0.0905
CB-2	0.6	0.3956	0.0489	0.1207	-0.0332	1.792	1.528	18.21	815.0	74.6	0.095	3.30	2.60	0.904	1.889	13.2376	0.3111
CB-3	0.8	0.3961	0.0883	0.1620	-0.0629	1.415	1.087	9.46	890.2	90.6	0.105	2.62	2.26	0.825	1.567	6.1533	0.6215

(CC-1 ~4)

$p_c = 7.383$MPa,	$T_c = 31.06$°C	$\rho_{Ls} = 658.0$kg/m³,	$\mu_{Ls} = 51.2\mu$Pas,	$\lambda_{Ls} = 0.079$W/mK,	$c_{PLs} = 10.48$kJ/kgK
$p_s = 6.8876$MPa,	$T_s = 28.0$°C	$\rho_{Vs} = 288.8$kg/m³,	$\mu_{Vs} = 22.5\mu$Pas,	$\lambda_{Vs} = 0.041$W/mK,	$c_{PVs} = 14.87$kJ/kgK

$\Delta h_V = 91.52$kJ/kg

$Pr_{Ls} = 6.81,$ $Pr_{Vs} = 8.17,$ $R_s = 2.27$

No.	η_{GLi}	F''_{GLw}	F_{GLi}	F'_{GLi}	F''_{GLi}	$-\Theta'_{GLi}$	$-\Theta'_{GLw}$	T_w °C	$\frac{\rho_{Lw}}{\text{kg/m}^3}$	$\frac{\mu_{Lw}}{\mu\text{Pas}}$	$\frac{\lambda_{Lw}}{\text{W/mK}}$	$\frac{c_{PLw}}{\text{kJ/kgK}}$	Pr_{Lw}	$\frac{q_{iz}}{q_{wz}}$	$\left(\frac{Nu_{Lwz}}{(GR/4)^{1/4}}\right)_r$	$\left(\frac{Pr_L}{H}\right)_r$	$\frac{H_L}{(\tau_T=0.9)}$
CC-1	0.4	0.3360	0.0181	0.0671	-0.0138	2.565	2.407	26.74	700.7	56.6	0.083	6.35	4.36	0.956	2.624	48.8185	0.1335
CC-2	0.6	0.3953	0.0483	0.1186	-0.0378	1.773	1.537	23.14	767.0	66.3	0.089	4.02	3.00	0.876	1.903	13.3589	0.4323
CC-3	0.7	0.4043	0.0676	0.1412	-0.0534	1.548	1.288	20.21	804.1	72.6	0.093	3.32	2.59	0.832	1.713	8.6132	0.6195
CC-4	1.0	0.3898	0.1349	0.1924	-0.1089	1.143	0.846	6.96	912.0	96.0	0.109	2.48	2.19	0.706	1.395	3.5020	1.2400

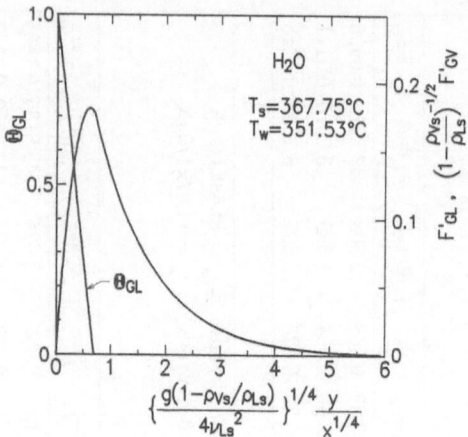

FIGURE 10.2-1. An example of the distributions of F'_{GL}, F'_{GV}, and Θ_{GL}.

Figure 10.2-2 shows the relation between $\{Nu_{Lwx}/(Gr_{Lx}/4)^{1/4}\}_r$ and $(Pr_L/H)_r$, where the former value is computed from

$$\left[\frac{Nu_{Lwx}}{\left(\dfrac{Gr_{Lx}}{4}\right)^{\frac{1}{4}}}\right]_r = (-\Theta'_{GLw}) \left(\frac{\rho_{Lw}\lambda_{Lw}}{\rho_{Lr}\lambda_{Lr}}\right) \left(\frac{\rho_{Lr}\mu_{Lr}}{\rho_{Ls}\mu_{Ls}}\right)^{\frac{1}{2}} \left(\frac{1-\dfrac{\rho_{Vs}}{\rho_{Ls}}}{1-\dfrac{\rho_{Vs}}{\rho_{Lr}}}\right)^{\frac{1}{4}}, \quad (10.2\text{-}26)$$

where

$$Gr_{Lx} = \frac{x^3 g \left(1 - \dfrac{\rho_{Vs}}{\rho_L}\right)}{\nu_L^2} \qquad (10.2\text{-}27)$$

and the representative physical properties are evaluated at $r_T = 1/2$. These data can be well correlated by the equation

$$(Nu_{Lwx})_r = \frac{1}{\sqrt{2}} \left(\frac{Gr_{Lx} Pr_L}{H}\right)^{\frac{1}{4}}_r. \qquad (10.2\text{-}28)$$

This equation coincides with Eq. (5.3-2) when ρ_{Vs} is much smaller than ρ_L.

Figure 10.2-3 shows the relation between q_{ix}/q_{wx} and H_r, in which c_{pL} is evaluated at $r_T = 0.9$. These data can be well correlated by Eq. (5.1-12), as seen in the figure.

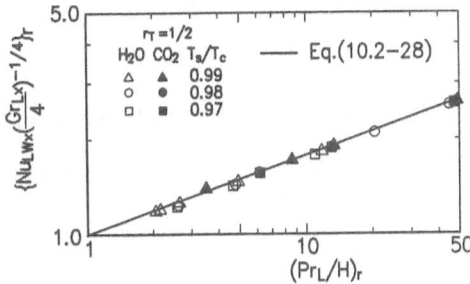

FIGURE 10.2-2. Correlation between $\{Nu_{Lwx}/(Gr_{Lx}/4)^{1/4}\}_r$ and $(Pr_L/H)_r$.

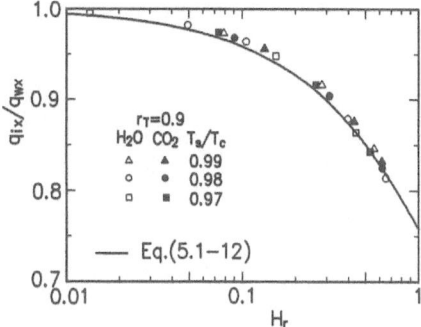

FIGURE 10.2-3. Correlation between q_{ix}/q_{wx} and H_r for free-convection condensation.

REFERENCES

1. Fujii, T., K. Shinzato, J.B. Lee, and M. Watabe (in preparation).

2. PROPATH-Group, PROPATH—A Program Package for Thermophysical Properties, Version 7.1, Corona, Tokyo (1990).

3. Fujii, T., K. Shinzato, J. B. Lee, and M. Watabe (in preparation).

Appendix:Physical Properties

The formulas of the physical properties of the seven substances used for computation in this book are presented as follows.

Substance		Molecular Weight M [kg/kmol]	Boiling Point T_b [K]	Critical Pressure p_c [MPa]	Critical Temperature T_c [K]
Air		28.97	78.8	3.766	132.5
Water	H_2O	18.015	373.15	22.01	647.13
Methanol	CH_3OH	32.042	337.8	8.10	512.58
Ethanol	C_2H_5OH	46.069	351.45	6.38	516.25
CFC11	CCl_3F	137.368	297.0	4.409	471.15
HCFC22	$CHClF_2$	86.469	232.3	4.988	369.3
CFC114	$CClF_2 \cdot CClF_2$	170.922	276.9	3.248	418.78

In this Appendix, the subscripts s and r mean saturation and reduced values, $\Re$ is the universal gas constant, and X is the molar fraction.

A1 Pressure p_s[kPa]-Temperature(T_s[K] or t_s[℃]) Relation at the Saturation State

Water:[1] $0 < t_s < 210℃$:

$$\log\left(\frac{p_s}{22.12 \times 10^3}\right) = -\left\{3.1323 + 3.116 \times 10^{-6}(210 - t_s)^{2.066}\right\}$$
$$\times \left(\frac{6.473}{273.15 + t_s} - 1\right). \tag{A1-1}$$

Methanol:[2] $-16 \le t_s \le 91℃$:

$$\log p_s = 7.19736 - \frac{1574.99}{(238.86 + t_s)}. \tag{A1-2}$$

Ethanol:[2] $-3 \leq t_s \leq 97°C$:

$$\log p_s = 7.33827 - \frac{1652.05}{(231.48 + t_s)} . \tag{A1-3}$$

The formulas for other substances are expressed by

$$\log p_r = -\frac{1}{a}\left(\frac{1}{T_r} - 1\right), \tag{A1-4}$$

where

$$p_r = \frac{p_s}{p_c}, \tag{A1-5a}$$

$$T_r = \frac{T_s}{T_c}, \tag{A1-5b}$$

$$A = \frac{1}{T_r} - 1 \tag{A1-5c}$$

and the values of a are given as follows:

CFC11:[3]

$$a = 0.350 + 0.0098 \, (1 + 7.8 \times 10^{-4} p_r^{-1.65})^{-\frac{2}{3}}$$
$$0.004 \leq p_r \leq 0.1 , \tag{A1-6a}$$

$$a = 0.350 + 0.0096 \, (1 + 90 p_r^4)^{-\frac{1}{2}}$$
$$0.1 \leq p_r \leq 0.2 , \tag{A1-6b}$$

$$\frac{1}{a} = 2.7 + 0.1 \, (6.5 \times 10^{-2} A^{-1.7} + 8 A^5)^{\frac{1}{4}}$$
$$0.54 \leq T_r \leq 0.78 . \tag{A1-6c}$$

HCFC22:[3]

$$a = 0.340 + 0.0120 \, (1 + 2 \times 10^{-3} p_r^{-1.6})^{-\frac{1}{2}}$$
$$0.007 \leq p_r \leq 0.1 , \tag{A1-7a}$$

$$a = 0.340 + 0.0116 \, (1 + 35 p_r^4)^{-\frac{1}{2}}$$
$$0.1 \leq p_r \leq 0.6 , \tag{A1-7b}$$

$$\frac{1}{a} = 2.8 + 0.1 \, (1.3 \times 10^{-3} A^{-\frac{8}{3}} + 7.3 A^{6.4})^{\frac{1}{4}}$$
$$0.57 \leq T_r \leq 0.94 . \tag{A1-7c}$$

CFC114:[4]

$$a = 0.320 + 0.0240 \, (1 + 1.6 \times 10^{10} - 2 p_r^{-1.2})^{-\frac{1}{4}}$$

$$0.002 \le p_r \le 0.1, \tag{A1-8a}$$

$$a = 0.320 + 0.0227 \left(1 + 253p_r^5\right)^{\frac{1}{4}}$$
$$0.1 \le p_r \le 0.5, \tag{A1-8b}$$

$$\frac{1}{a} = 2.8 + 0.1 \left(0.125A^{-1.8} + 16.7A^{2.9}\right)^{\frac{1}{2}}$$
$$0.5 \le T_r \le 0.9. \tag{A1-8c}$$

A2 Phase Equilibrium

The liquid and vapor lines in a phase equilibrium diagram can be derived by the equation

$$X_{kL} = \frac{p}{\gamma_k p_k^0(T)} X_{kV} \qquad (k = 1, 2, \ldots, n), \tag{A2-1}$$

where X_{kL} and X_{kV} are the molar fractions of the liquid and vapor of component k, respectively. p is the total pressure, $p_k^0(T)$ is the saturation pressure of component k, and γ_k is the activity constant. γ_k are usually taken as

$$\gamma_1 = \gamma_2 = 1, \tag{A2-2}$$

particularly for an ethanol-water mixture:[5]

$$\ln\gamma_1 = \frac{2187.6X_{2L}^2 - 2407.2X_{2L}^3 + 5398.7X_{2L}^4}{\Re T}, \tag{A2-3a}$$

$$\ln\gamma_2 = \frac{9374.3X_{1L}^2 - 11989.4X_{1L}^3 + 5398.7X_{2L}^4}{\Re T} \tag{A2-3b}$$

and for methanol-ethanol-water:[6]

$$\log\gamma_1 = \frac{1}{T}\left(\frac{6.85X_{3L}}{X_{1L} + 1.275X_{2L} + 0.62X_{3L}}\right)^2, \tag{A2-4a}$$

$$\log\gamma_2 = \frac{1}{T}\left(\frac{9.05X_{3L}}{0.885X_{1L} + X_{2L} + 0.55X_{3L}}\right)^2, \tag{A2-4b}$$

$$\log\gamma_3 = \frac{1}{T}\left(\frac{1.40X_{1L} + 22.3X_{2L}}{1.61X_{1L} + 1.82X_{2L} + X_{3L}}\right)^2. \tag{A2-4c}$$

A3 Density ρ [kg/m^3]

The density of gases and vapors is expressed as

$$\rho_V = \frac{Mp}{Z\Re T}, \tag{A3-1}$$

where p[kPa] is the pressure, T[K] is the temperature, M[kg/kmol] is the molecular weight, $\Re = 8.3145$[kJ/kmol K] is the universal gas constant, and Z is the compressibility factor.

Air:[1] $270 \leq T \leq 570$ K, $0 < p < 2$ MPa :

$$Z = 1, \qquad \rho_V = 3.483 \frac{p}{T} \,. \tag{A3-2}$$

Water:[1]

Saturated vapor: $273.15 < T < 543$ K :

$$\frac{1}{Z_s} = 1 + 0.0422 \left(\frac{p}{T}\right)^{0.8} \,. \tag{A3-3}$$

Superheated vapor: $T - T_s < 150$ K :

$$\frac{Z}{Z_s} = 1 + 3.6 \times 10^{-5} (T - T_s)^{0.365} p^{0.7} \log (T - T_s + 1). \tag{A3-4}$$

Saturated and subcooled liquid: $0 \leq t \leq 220$°C :

$$\frac{1}{\rho_L} = (1 + 8.7 \times 10^{-6} t^{1.85}) \times 10^{-3}. \tag{A3-5}$$

Methanol:[7]

Vapor: $338.15 \leq T \leq 373.15$ K :

$$Z = 1, \qquad \rho_V = 3.854 \frac{p}{T} \,. \tag{A3-6}$$

Liquid: $65 \leq t \leq 100$°C :

$$\rho_L = 812 - 0.98t \,. \tag{A3-7}$$

Ethanol:[5]
Vapor: $273.15 < T < 373.15$ K :

$$\frac{1}{Z_s} = 1 + 0.99 \left(\frac{p}{T}\right). \tag{A3-8}$$

Liquid: $0 \leq t \leq 100$°C :

$$\rho_L = -0.9055t + 807.44 \,. \tag{A3-9}$$

CFC11:[3]
Saturated vapor: $0.004 \leq p_r \leq 0.2$:

$$\frac{1}{Z_s} = 1 + 0.630 \left(p_r^{0.726} + p_r^{2.71}\right). \tag{A3-10}$$

Superheated vapor: $0.0365 \leq p_r \leq 0.6$:

$$\frac{Z}{Z_s} = 1 + 8\log\left(1 + \Delta T_r\right)(1 + \Delta T_r)^{-2.7}(p_r^{0.72} + p_r^{5.2}), \quad \text{(A3-11)}$$

where

$$\Delta T_r = \frac{T - T_s}{T_c} . \qquad \text{(A3-12)}$$

Liquid: $0 \leq t \leq 70°C$:

$$\frac{1}{\rho_L} = (0.652 + 7.52 \times 10^{-4}t^{1.1}) \times 10^{-3}. \qquad \text{(A3-13)}$$

HCFC22:[3]

Saturated vapor: $0.007 \leq p_r \leq 0.6$:

$$\frac{1}{Z_s} = 1 + 0.650\,(p_r^{0.726} + p_r^{2.71}) . \qquad \text{(A3-14)}$$

Superheated vapor: $0.03 \leq p_r \leq 0.6$:

$$\frac{Z}{Z_s} = 1 + 8\log\left(1 + \Delta T_r\right)(1 + \Delta T_r)^{-2.7}(p_r^{0.72} + p_r^{5.2}). \qquad \text{(A3-15)}$$

Liquid: $-60 \leq t \leq 70°C$:

$$\frac{1}{\rho_L} = \{0.766 + 1.37 \times 10^{-3}t + 2.16 \times 10^{-8}(t + 60)^{3.25}\} \times 10^{-3} .$$
$$\text{(A3-16)}$$

CFC114:[4]

Saturated vapor: $0.03 \leq p_r \leq 0.5$:

$$\frac{1}{Z_s} = 1 + 0.634\,(p_r^{0.785} + p_r^{3.03}) . \qquad \text{(A3-17)}$$

Superheated vapor: $0.03 \leq p_r \leq 0.5$:

$$\begin{aligned}\frac{Z}{Z_s} &= 1 + 8\log\left(1 + \Delta T_r\right)(1 + \Delta T_r)^{-2.7} \\ &\quad \times (0.73p_r^{0.72} + 1.73p_r^{2.8}) .\end{aligned} \qquad \text{(A3-18)}$$

Liquid: $-60 \leq t \leq 70°C$:

$$\frac{1}{\rho_L} = \{0.647 + 8.81 \times 10^{-4}t + 4.35 \times 10^{-8}(t + 60)^{2.9}\} \times 10^{-3}.$$
$$\text{(A3-19)}$$

A4   Isobaric Specific Heat c_p [J/kg K]

Air:[1] $270 < T < 570$ K :

$$c_{pV} = (1 + 2.5 \times 10^{-10}T^3) \times 10^3 . \tag{A4-1}$$

Water:[1]
Saturated vapor: $0 \le t \le 290°C$:

$$c_{pV} = 1.863 \times 10^3 + 1.65 \times 10^{-3}t^{2.5} + 1.2 \times 10^{-18}t^{8.5} . \tag{A4-2}$$

Liquid: $10 \le t \le 230°C$:

$$c_{pL} = 4.179 \times 10^3 + 7.9 \times 10^{-5}(t - 10)^{2.9} . \tag{A4-3}$$

Methanol:
Saturated vapor:[8] $0 \le t \le 100°C$:

$$c_{pV} = 1.34 \times 10^3 + 2.6t + 0.14t^{1.4} . \tag{A4-4}$$

Liquid:[9]

$$c_{pL} = 2.43 \times 10^3 + 2.3t , \quad 0 \le t \le 50°C , \tag{A4-5}$$
$$c_{pL} = 1.96 \times 10^3 + 11.7t , \quad 50 \le t \le 100°C . \tag{A4-6}$$

Ethanol:[5]
Saturated vapor: $0 \le t \le 150°C$:

$$c_{pV} = 1.52 \times 10^3 + 2.9t^{1.011} . \tag{A4-7}$$

Liquid: $0 \le t \le 100°C$:

$$c_{pL} = 2.262 \times 10^3 + 6.53t + 0.094t^{1.79} . \tag{A4-8}$$

CFC11:[3]
Saturated vapor: $0.004 \le p_r \le 0.2$:

$$\frac{(c_{pV} - c_{p0})}{T} = 0.76p_r^{0.61} + 4.46p_r^{3.24} , \tag{A4-9}$$

where

$$c_{p0} = (0.545 + 8.13 \times 10^{-4}t) \times 10^3, \quad -20 \le t \le 60°C , \tag{A4-10}$$
$$c_{p0} = (0.558 + 6.03 \times 10^{-4}t) \times 10^3, \quad 60 \le t \le 150°C . \tag{A4-11}$$

Liquid: $0 \le t \le 70°C$:

$$c_{pL} = (0.867 + 8.10 \times 10^{-4}t + 8.10 \times 10^{-9}t^{3.6}) \times 10^3 . \tag{A4-12}$$

HCFC22:[3]

Saturated vapor: $0.007 \leq p_r \leq 0.6$:

$$\frac{(c_{pV} - c_{p0})}{T} = 1.90 p_r^{0.61} + 6.15 p_r^{3.24} , \qquad (A4\text{-}13)$$

where

$$c_{p0} = (0.615 + 1.19 \times 10^{-3}t) \times 10^3, \quad -60 \leq t \leq 90°C . \qquad (A4\text{-}14)$$

Liquid: $-60 \leq t \leq 70°C$:

$$c_{pL} = \{1.142 + 1.13 \times 10^{-3}t + 2.25 \times 10^{-8}(t + 60)^{3.41}\} \times 10^3. \qquad (A4\text{-}15)$$

CFC114:[4]

Saturated vapor: $0.03 \leq p_r \leq 0.5$:

$$\frac{(c_{pV} - c_{p0})}{T} = 0.40 p_r^{0.61} + 2.75 p_r^{3.24} , \qquad (A4\text{-}16)$$

where

$$c_{p0} = (0.674 + 1.22 \times 10^{-3}t) \times 10^3, \quad 0 \leq t \leq 50°C , \qquad (A4\text{-}17)$$
$$c_{p0} = (0.685 + 1.00 \times 10^{-3}t) \times 10^3, \quad 50 \leq t \leq 140°C . \qquad (A4\text{-}18)$$

Liquid: $-60 \leq t \leq 70°C$:

$$c_{pL} = (0.955 + 2.40 \times 10^{-3}t) \times 10^3 . \qquad (A4\text{-}19)$$

A5 Latent Heat Δh_V of Condensation and Enthalpy h [kJ/kg]

Latent heat of condensation Δh_V for a pure substance is given by the enthalpy difference between vapor and liquid at the saturation state.

Water:[9] $0 \leq t \leq 220°C$:

$$h_V = 2501.6 + 1.81t^{-6.68} \times 10^{-7}t^{3.5} , \qquad (A5\text{-}1)$$
$$h_L = 4.2t + 7.5 \times 10^{-9}t^4 . \qquad (A5\text{-}2)$$

Methanol:[9] $0 \leq t \leq 150°C$:

$$h_V = 1200.0 + 1.224t^{-0.311} \times 10^{-2}t^2 , \qquad (A5\text{-}3)$$
$$h_L = 2.40t + 3.10 \times 10^{-4}t^{2.5} . \qquad (A5\text{-}4)$$

Ethanol:[9] $0 \le t \le 180°C$:

$$h_V = 946.5 + 1.247t , \tag{A5-5}$$
$$h_L = 2.28t + 3.687 \times 10^{-4}t^{2.5} . \tag{A5-6}$$

CFC11:[9] $210 \le T \le 460$ K :

$$h_V = 252.3 + 0.5T - 10^{(0.0127T - 4.365)} , \tag{A5-7}$$
$$h_L = -33.73 + 0.855T + 10^{(0.0119T - 3.96)} . \tag{A5-8}$$

HCFC22:[9] $170 \le T \le 360$ K :

$$h_V = 275.8 + 0.4825T - 10^{(0.01282T - 2.971)} , \tag{A5-9}$$
$$h_L = -113.0 + 1.147T + 10^{(0.02175T - 6.358)} . \tag{A5-10}$$

CFC114:[9] $200 \le T \le 410$ K :

$$h_V = 364.1 - 1.475T + 7.487 \times 10^{-3}T^2 - 8.959 \times 10^{-6}T^3, \tag{A5-11}$$
$$h_L = 42.86 + 0.2357T + 1.243 \times 10^{-3}T^2 . \tag{A5-12}$$

A6   Thermal Conductivity λ [W/mK]

Air:[1] $270 \le T \le 570$ K :

$$\lambda_V = \frac{1.195 \times 10^{-3}T^{1.6}}{118 + T} . \tag{A6-1}$$

Water:[1]

Saturated vapor: $0 \le t \le 290°C$:

$$\lambda_V = (1.87 + 1.65 \times 10^{-3}t^{\frac{9}{7}} + 5.7 \times 10^{-13}t^{5.1}) \times 10^{-2}. \tag{A6-2}$$

Liquid: $10 \le t \le 135°C$:

$$\lambda_L = 0.6881 - 4 \times 10^{-6}(135 - t)^{2.1} . \tag{A6-3}$$

Methanol:[7]

Saturated vapor: $65 \le t \le 100°C$:

$$\lambda_V = 0.0116 + 1.4 \times 10^{-4}t . \tag{A6-4}$$

Liquid: $65 \le t \le 100°C$:

$$\lambda_L = 0.208 - 3.0 \times 10^{-4}t . \tag{A6-5}$$

Ethanol:[5]
Saturated vapor: $0 \leq t \leq 150°C$:

$$\lambda_V = 0.01257 + 9.94 \times 10^{-5}t. \tag{A6-6}$$

Liquid: $0 \leq t \leq 100°C$:

$$\lambda_L = 0.17256 - 2.3412 \times 10^{-4}t. \tag{A6-7}$$

CFC11:[3]
Saturated vapor: $20 \leq t \leq 150°C$:

$$\lambda_V = (6.542 + 4.77 \times 10^{-2}t) \times 10^{-3}. \tag{A6-8}$$

Liquid: $0 \leq t \leq 70°C$:

$$\lambda_L = 0.0943 - 2.75 \times 10^{-4}t. \tag{A6-9}$$

HCFC22:[3]
Saturated vapor: $-40 \leq t \leq 90°C$:

$$\lambda_V = \{9.400 + 6.00 \times 10^{-2}t + 2.70 \times 10^{-27}(t+40)^{13}\} \times 10^{-3}. \tag{A6-10}$$

Liquid: $-60 \leq t \leq 70°C$:

$$\lambda_L = 0.0988 - 4.63 \times 10^{-4}t^{-5.48} \times 10^{-16}(t+60)^{6.3}. \tag{A6-11}$$

CFC114:[4]
Saturated vapor: $0 \leq t \leq 120°C$:

$$\lambda_V = (9.45 + 6.5 \times 10^{-2}t + 3.7 \times 10^{-9}t^{3.8}) \times 10^{-3}. \tag{A6-12}$$

Liquid: $-30 \leq t \leq 110°C$:

$$\lambda_L = 0.0710 - 2.67 \times 10^{-4}t. \tag{A6-13}$$

A7 Viscosity μ [kg/ms] (T[K], t[°C])

Air:[1] $270 \leq T \leq 570$ K :

$$\mu_V = \frac{1.488 \times 10^{-6}T^{\frac{3}{2}}}{118 + T}. \tag{A7-1}$$

Water:[1]
Saturated vapor: $0 \leq t \leq 100°C$:

$$\mu_V = (8.02 + 0.04t) \times 10^{-6} . \tag{A7-2}$$

Liquid: $0 \leq t \leq 300°C$:

$$\mu_L = 2.4 \times 10^{-5} \times 10^{\frac{251}{t+135}} . \tag{A7-3}$$

Methanol:[7]
Saturated vapor: $65 \leq t \leq 100°C$:

$$\mu_V = (8.8 + 0.034t) \times 10^{-6} . \tag{A7-4}$$

Liquid: $65 \leq t \leq 100°C$:

$$\mu_L = 7.161 \times 10^{-4} \times 10^{(-0.00528t)} . \tag{A7-5}$$

Ethanol:[4]
Saturated vapor: $0 \leq t \leq 50°C$:

$$\mu_V = (76.33 + 0.33425t) \times 10^{-7} . \tag{A7-6}$$

Liquid: $0 \leq t \leq 100°C$:

$$\mu_L = 1.545 \times 10^{-7} \times 10^{\frac{1817}{t+447.22}} . \tag{A7-7}$$

CFC11:[3]
Saturated vapor: $10 \leq t \leq 150°C$:

$$\mu_V = (0.990 + 4.00 \times 10^{-3}t + 8.00 \times 10^{-14} t^{5.4}) \times 10^{-5}. \tag{A7-8}$$

Liquid: $0 \leq t \leq 70°C$:

$$\mu_L = 2.29 \times 10^{-5} \times 10^{\frac{385}{t+281}} . \tag{A7-9}$$

HCFC22:[3]
Saturated vapor: $-40 \leq t \leq 80°C$:

$$\mu_V = \{(1.184 + 4.60 \times 10^{-3}t + 3.65 \times 10^{-10}(t + 40)^{4.27})\} \times 10^{-5}. \tag{A7-10}$$

Liquid: $0 \leq t \leq 70°C$:

$$\mu_L = 2.20 \times 10^{-5} \times 10^{\frac{326}{t+318}} . \tag{A7-11}$$

CFC114:[4]

Saturated vapor: $-30 \leq t \leq 120°C$:

$$\mu_V = \{1.08 + 3.25 \times 10^{-3}t + 1.50 \times 10^{-11}(t + 30)^{4.7}\} \times 10^{-5}. \tag{A7-12}$$

Liquid: $-50 \leq t \leq 70°C$:

$$\mu_L = 1.50 \times 10^{-5} \times 10^{\frac{418.5}{t + 277}}. \tag{A7-13}$$

A8 Diffusivity of Gas Phase D $[cm^2/s](T[K],$ $p[kPa])$

Air-water:[1] $270 \leq T \leq 570$ K :

$$D = \frac{7.65 \times 10^{-4}T^{\frac{11}{6}}}{p}. \tag{A8-1}$$

Air-Ethanol:[5]

$$D = \frac{3.62 \times 10^{-4}T^{1.83}}{p}. \tag{A8-2}$$

Ethanol-Water:[5]

$$D = \frac{4.58 \times 10^{-4}T^{1.83}}{p}. \tag{A8-3}$$

The prediction formula[2] for other binary vapor mixtures is ;

$$D = \frac{6.61 \times 10^{-6}T^{1.83}}{p\left\{\left(\frac{T_c}{p_c}\right)_1^{\frac{1}{3}} + \left(\frac{T_c}{p_c}\right)_2^{\frac{1}{3}}\right\}^3}\sqrt{\frac{1}{M_1} + \frac{1}{M_2}}. \tag{A8-4}$$

A9 Equations for Predicting Physical Properties of Mixtures

Latent heat:

$$\Delta h_V \approx \sum_{k=1}^{n} W_{kL}\Delta h_k, \tag{A9-1}$$

where $\Delta h_k = h_V - h_L[kJ/kg]$ is the latent heat of component k.

Density of vapors:

$$\rho_V = \sum_{k=1}^{n} \rho_{kV} \,. \qquad (A9\text{-}2)$$

Density of liquids:

$$\frac{1}{\rho_L} = \sum_{k=1}^{n} \frac{W_{kL}}{\rho_{kL}} \,. \qquad (A9\text{-}3)$$

Isobaric specific heat of vapors or liquids:

$$c_p = \sum_{k=1}^{n} W_k c_{pk} \,. \qquad (A9\text{-}4)$$

Thermal conductivity of vapors:

$$\lambda_V = \sum_{k=1}^{n} \frac{X_{kV} \lambda_{kv}}{\sum\limits_{l=1}^{n} X_{lV} A_{kl}} \,, \qquad (A9\text{-}5)$$

where

$$A_{kl} = \frac{1}{4}\left[1 + \left\{\frac{\mu_{kV}}{\mu_{lV}}\left(\frac{M_l}{M_k}\right)^{\frac{3}{4}}\frac{T+S_k}{T+S_l}\right\}^{\frac{1}{2}}\right]^2 \frac{T+S_{kl}}{T+S_k} \,, \qquad (A9\text{-}6)$$

$$S_k = 1.5 T_{kb} \,, \qquad (A9\text{-}7a)$$

$$S_{kl} = 0.733 \, (S_k S_l)^{\frac{1}{2}} \,. \qquad (A9\text{-}7b)$$

Thermal conductivity of liquids:[10,11]

$$\lambda_L = \sum_{k=1}^{n}\sum_{l=1}^{n} y_{kl} y_{lL} \lambda_{klL} \,, \qquad (A9\text{-}8)$$

where

$$\frac{2}{\lambda_{klL}} = \frac{1}{\lambda_{kL}} + \frac{1}{\lambda_{lL}} \,, \qquad (A9\text{-}9)$$

$$y_{kL} = W_{kL}\frac{\rho_{kL}}{\rho_L} \,. \qquad (A9\text{-}10)$$

Viscosity of vapors:

$$\mu_V = \sum_{k=1}^{n} \frac{X_{kV}\mu_{kv}}{\sum_{l=1}^{n} X_{lV}\phi_{kl}}, \qquad \text{(A9-11)}$$

where

$$\phi_{kl} = \frac{\left\{1+\left(\frac{\mu_{kV}}{\mu_{lV}}\right)^{\frac{1}{2}}\left(\frac{M_l}{M_k}\right)^{\frac{1}{4}}\right\}^2}{\left\{8\left(1+\frac{M_k}{M_l}\right)\right\}^{\frac{1}{2}}}. \qquad \text{(A9-12)}$$

Viscosity of liquids consisting of nonpolar molecules:

$$\ln \mu_L = \sum_{k=1}^{n} X_{kL} \ln \mu_{kL}. \qquad \text{(A9-13)}$$

A10   Viscosity of Mixtures of Polar Molecules

This value can be predicted by Panchenkov's[12] formula :

$$\mu_L = k\rho_L^{\frac{4}{3}}T^{\frac{1}{2}}\left\{\exp\left(\frac{E}{\Re T}\right)-1\right\}; \qquad \text{(A10-1)}$$

μ_L [kg/ms], ρ_L [kg/m³], T [K], k[–], and E [kJ/kmol].

Methanol-Water:

$$k = \left\{\exp\left(-34.39X_{1L}^4 + 12.59X_{1L}^3 + 21.19X_{1L}^2 - 9.811X_{1L}\right.\right.$$
$$\left.\left.-0.3169\right)\right\} \times 10^{-11}, \quad 0 \leq X_{1L} < 0.482; \qquad \text{(A10-2a)}$$

$$k = \left\{\exp\left(22.88X_{1L}^4 - 60.07X_{1L}^3 + 58.71X_{1L}^2 - 20.99X_{1L}\right.\right.$$
$$\left.\left.-1.40\right)\right\} \times 10^{-11}, \quad 0.482 \leq X_{1L} \leq 1.0; \qquad \text{(A10-2b)}$$

$$E = \left\{\exp\left(-3.713X_{1L}^4 + 8.503X_{1L}^3 - 7.616X_{1L}^2 + 2.370X_{1L}\right.\right.$$
$$\left.\left.+0.4861\right)\right\} \times 10^4. \qquad \text{(A10-3)}$$

Ethanol-Water:

$$k = \left[\exp\left\{4066(X_{1L} - 0.2)^4 + 1275(X_{1L} - 0.2)^3\right.\right.$$
$$\left.+187.2(X_{1L} - 0.2)^2 + 7.894(X_{1L} - 0.2) - 2.527\right\}\right] \times 10^{-11},$$
$$0.0 \leq X_{1L} \leq 0.2; \tag{A10-4a}$$

$$k = \left\{\exp\left(-12.92X_{1L}^4 + 33.43X_{1L}^3 - 31.59X_{1L}^2 + 16.93X_{1L}\right.\right.$$
$$\left.\left.-4.897\right)\right\} \times 10^{-11}, \quad 0.2 \leq X_{1L} \leq 1.0; \tag{A10-4b}$$

$$E = \left\{\exp\left(-545.5X_{1L} + 291.8X_{1L} - 65.10X_{1L} + 7.758X_{1L}\right.\right.$$
$$\left.\left.+0.4869\right)\right\} \times 10^4, \quad 0.0 \leq X_{1L} \leq 0.2; \tag{A10-5a}$$

$$E = \left\{\exp\left(1.168X_{1L} - 2.922X_{1L} + 2.422X_{1L} - 1.4015X_{1L}\right.\right.$$
$$\left.\left.+1.0994\right)\right\} \times 10^4, \quad 0.2 < X_{1L} \leq 1.0. \tag{A10-5b}$$

Methanol-Ethanol:

$$k = \left(-1.981X_{1L}^3 + 3.936X_{1L}^2 + 2.326X_{1L} + 2.644\right) \times 10^{-11}, \tag{A10-6}$$

$$E = (-0.4166X_{1L} + 1.439) \times 10^4. \tag{A10-7}$$

Methanol-Ethanol-Water:

Watabe[9] proposed the equation

$$k = \left\{\frac{X_{1L}}{X_{1L} + X_{2L}}k(X_{1L} + X_{2L})_{13} + \frac{X_{2L}}{X_{1L} + X_{2L}}k(X_{1L} + X_{2L})_{23}\right\}$$
$$\times \left[1 + (X_{1L} + X_{2L})\left\{\frac{k\left(\dfrac{X_{1L}}{X_{1L} + X_{2L}}\right)_{12}}{k^*_{12}} - 1\right\}\right], \tag{A10-8}$$

$$E = \left\{\frac{X_{1L}}{X_{1L} + X_{2L}}E(X_{1L} + X_{2L})_{13} + \frac{X_{2L}}{X_{1L} + X_{2L}}E(X_{1L} + X_{2L})_{23}\right\}$$
$$\times \left[1 + (X_{1L} + X_{2L})\left\{\frac{E\left(\dfrac{X_{1L}}{X_{1L} + X_{2L}}\right)_{12}}{E^*_{12}} - 1\right\}\right], \tag{A10-9}$$

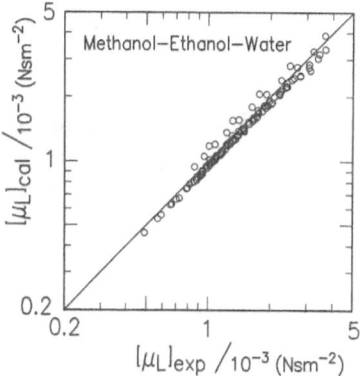

FIGURE A10-1. Comparison in the viscosity of a methanol-ethanol-water liquid mixture between the values measured by Dizechi and Marschall (Ref. 10) and the values calculated from Eqs. (A10-1), (A10-8), and (A10-9).

where $k(X)_{ij}$ and $E(X)_{ij}$ are the values at the molecular fracton X for a binary mixture of i and j components, and

$$k^*_{12} = \frac{X_{1L}}{X_{1L} + X_{2L}} k_1 + \frac{X_{2L}}{X_{1L} + X_{2L}} k_2 , \qquad \text{(A10-10)}$$

$$E^*_{12} = \frac{X_{1L}}{X_{1L} + X_{2L}} E_1 + \frac{X_{2L}}{X_{1L} + X_{2L}} E_2 . \qquad \text{(A10-11)}$$

These equations can correlate the data measured by Dizechi and Marshall[13] within an error on 10 percent, as shown in Fig. A10-1.

REFERENCES

1. Fujii, T., Y. Kato, and K. Mihara, Expressions of Transport and Thermodynamic Properties of Air, Steam and Water (in Japanese), *Reports of Research Institute of Industrial Science, Kyushu University*, 66, 81–95 (1977).

2. Kagaku-Kogaku Kyokai, "Kagaku-Kogaku Handbook," 5th Ed. (in Japanese), Maruzen, Tokyo (1988).

3. Fujii, T., Sh. Nozu, and H. Honda, Expressions of Thermodynamic and Transport Properties of Refrigerants R11, R12, R22 and R113 (in Japanese), *Reports of Research Institute of Industrial Science, Kyushu University*, 67, 43–59 (1987).

4. Nozu, Sh., Personal communication.

5. Fujii, T., Sh. Koyama, and Sh. Nishida, Expression of Physical Properties Concerning Convection of Ethanol-Water-Air Mixture (in Japanese), *Reports of Research Institute of Industrial Science, Kyushu University*, 75, 63–76 (1983).

6. Kogan, V. B., V. M. Fridman, and V. V. Kafarov, "Equilibrium between Liquid and Vapor" (in Russian), Nauka, Moscow (1966).

7. Fujii, T. and Y. Kato, Laminar Film Condensation of a Binary Vapour on a Flat Surface (in Japanese), *Trans. Jpn. Soc. Mech. Eng.*, 46, 402, B, 306–312 (1980).

8. Maddox, R. N., Properties of saturated fluids, " Heat Exchanger Deisign Handbook," pp. 5.5.1–17, Hemisphere, New York (1983).

9. Watabe, M., Theoretical Study of Condensation of a Multicomponent Vapor Mixture of a Flat Plate (in Japanese), *Doctoral Thesis, Kyushu University* (1989).

10. Chen, Ze-Shao, T. Fujii, M. Fujii and Xin-Shi Ge, An Equation of Binary Liquid Mixtures (in Japanese), *Reports of Research Institute of Industrial Science, Kyushu University*, 82, 159–171 (1987).

11. Chen, Ze-Shao, T. Fujii and M. Fujii, A Method for Evaluating the Mixture Constant in the Expression for Thermal Conductivity of Binary Liquid Mixtures and its Extension to Multicomponent Liquid Mixtures (in Japanese), *Reports of Research Institute of Industrial Science, Kyushu University*, 82, 173–187 (1987).

12. Kagaku-Kogaku Kyokai, "Physical Properties" (in Japanese), Vol. 1-10, Maruzen, Tokyo (1963-1973).

13. Dizechi, M. and E. Marshall, Viscosity of Some Binary and Ternary Liquid Mixtures, *J. Chem. Eng. Data*, 27, 358 (1982).

14. JSME Data Book, "Thermophysical Properties of Fluid," Japan Society of Mechanical Engineers (1983).

Author Index

Subject Index